"*The Grandest Challenge* shows it's possible to bridge the global health gap between rich and poor using innovation." — *The Gazette* (Montreal)

"*The Grandest Challenge* is not only enlightening, solution orientated and deeply personal but it also encourages the reader to challenge the existing norm and encourages us to ask ourselves pivotal questions."
 — *The Independent* (UK)

"An engaging account . . . inspirational. The authors of this book are role models for how empathy and dedication to social justice can serve as powerful forces to solve our most difficult global challenges."
 — Dr. Calestous Juma, Kennedy School, Harvard University

"Daar and Singer show us how to rise to the grandest challenge of all: to make sure that science keeps its promise to all people, not just the rich. They point the way to how we can heal the sick, protect the vulnerable, reach the unreached, and bring human dignity to all. Essential reading."
 — Ismail Serageldin, Director of the Bibliotheca Alexandrina

"Inspired and inspiring. What struck me most when reading *The Grandest Challenge* was the perseverance, determination and patience of those endeavouring to improve the desperate and, in many cases, life-threatening situations in the developing world. Hope, positive thinking and optimism are always evident in Daar and Singer's admirable book."
 — Judith Razek, "Words in Action," Radio Oman FM

"[A] vivid and well-written book. . . . *The Grandest Challenge* is an accessible, informative and valuable resource." — *The Catholic Register*

DR. ABDALLAH DAAR & DR. PETER SINGER

THE GRANDEST CHALLENGE

Bringing Life-Saving Science from Lab to Village

Edited by Sarah Scott

ANCHOR CANADA

**Library and Archives Canada Cataloguing in Publication
is available upon request**

ISBN 978-0-385-66719-7

Cover and text design: Andrew Roberts

Printed and bound in the USA

Published in Canada by Anchor Canada,
a division of Random House of Canada Limited

Visit Random House of Canada Limited's website: www.randomhouse.ca

10 9 8 7 6 5 4 3 2 1

To our wives, Heather and Shahina;
our children, Lamees, Nadia, Erin, Rebecca, David, and Marwan;
and to the billions who dream of a better and more equal life.

CONTENTS

PROLOGUE

Our inspiration for writing this book stems from a simple but powerful fact: If your home is in London, New York or Toronto, you can expect to live well into your late seventies or early eighties. If you live in Morogoro, Tanzania, on the other hand, you will live only about half as long.

In fact, roughly 90 percent[1] of human beings live in poor regions of the world, and their lives are routinely cut short by infectious diseases such as malaria and tuberculosis and HIV/AIDS, diseases that in rich countries are preventable or controllable. The overwhelming majority of the world's citizens are also increasingly more susceptible to non-infectious diseases, such as diabetes and heart disease, which in developing nations kill more readily because of poor diet, poor living conditions and limited access to health care.[2]

For many decades and perhaps even longer, most of us have accepted these inequities as inevitable, a function of the great divide between the lesser-developed world and prosperous nations. But does it really have to be this way? Is there any solid evidence to suggest that life expectancy in some places must necessarily be shorter than in others? And is a life in the developed world really more valuable than the life of someone living in a poorer region? Absolutely not.

In fact, with the mapping of the human genome we are poised at the edge of a revolutionary wave of science, one that offers incredible opportunities to improve life sciences and medical possibilities—for all people on Earth. In this book, we will tour the groundbreaking new discoveries and technologies spawned by the Human Genome Project, innovations that will change many aspects of modern medicine, including the way drugs and vaccines are discovered and developed, and how diagnoses are made.

We will also pose a critical question: now that we have such extraordinary capabilities, what will we use this power to achieve? To create designer drugs for the rich, which will only widen the health gap between rich and poor? Or will we finally choose to tap the enormous potential of the new science to address the needs of 90 percent of the world?

We are two scientists—one born in an affluent Canadian city, the other in a poor Tanzanian one. A number of years ago, we joined forces to pursue an audacious goal: to use modern genomic science to give a child in sub-Saharan Africa—or any other disadvantaged area—the chance to live as long and healthy a life as a child in New York or Toronto. Contrary to what many people believe, this goal is not a grandiose, abstract ambition. Rather, it is a scientific and ethical mission that, even as we write these words, is being put into practice in labs and in villages around the globe.

It is true that until recently, the Human Genome Project, an epic scientific achievement that is turning information hidden in the double helix of DNA into a readable language, ignored the needs of the developing world. Most scientists thought that genomic research and applications were too sophisticated to be of use to people in poor countries. But imagine if we challenged that thinking. Imagine what would happen if the ability to read the code of life were applied specifically to problems of the poor. Suddenly, recently discovered cutting-edge powers to heal would be available to the very people who need them most. This kind of thinking could positively change the world we all live in, and save tens of millions of lives.

Here's the good news: it's now starting to happen. Exciting research is under way all around the world, and in this book we will be your guides through both the thrilling scientific and medical exploits of the past decade and the most promising of today's innovations in life sciences, those that hold the greatest possibility for improving the health of the world's poorest and most disenfranchised citizens. We will tell the story of how, in 2003, a half-a-billion-dollar investment from the Bill & Melinda Gates Foundation was deployed to help researchers use the new life-sciences revolution to fight ancient diseases such as malaria from new angles. In California, Australia and London, scientists are now tweaking the DNA of mosquitoes to stop them from spreading malaria. In Seattle, a malaria vaccine is being developed that knocks out genes the malaria parasite needs to survive in the human body.[3] Some of the world's greatest researchers are, at last, focusing on the problem of malaria in the developing world and using the newest scientific breakthroughs to address an age-old scourge. And this scientific awakening is not limited just to malaria but is growing to cover problems as diverse as dengue fever, tuberculosis, pneumonia and diarrhea, all of which are serious ailments in the developing world.

But acquiring the knowledge is only one half of the solution, and only one half of the story we tell. How do you transfer the knowledge found in labs around the world right to the hands of the poor in disadvantaged villages? Both of us have devoted much of our professional lives to answering that very question, and what we can tell you is that it's a long road from lab to village, with many roadblocks to overcome. In this book, we will take you to labs in India and China that have become potent new players in biotechnology and in the discovery and manufacture of drugs. We will go to Bangalore, where one of India's richest women started a biotechnology company that supplies affordable insulin to India's growing population of diabetics. We will go to Hyderabad, where a biotechnology company brought down the price of hepatitis B vaccine more than twentyfold through clever thinking and innovation. In these pages, you will also meet scientists, thinkers, doctors, heads of state, CEOs, ethicists and ordinary people who have taken up the call to action to close the gap between the developed and developing worlds.

Beyond encountering the people and places on the forefront of a revolution in life sciences, you will also learn about the profound ethical challenges we face on the road ahead. How do you persuade a rural, isolated community to test a new genetically modified mosquito? How do you get female sex workers to try out an anti-HIV gel that could save their lives? How do you obtain consent to trials in a way that respects the health, cultural beliefs and social structures of individual traditional communities around the world? How do you overcome decades of suspicion of multinational companies so that private organizations can contribute to finding a cure to a killer disease? How do you promote the building of labs and other facilities, not in the developed world but close to the people who need them most? How do you encourage companies to make life-saving drugs and vaccines more affordable? And finally, how do you handle the political roadblocks that can stymie even the most creative and important of scientific endeavours?

We contend that if we can answer these questions and overcome these obstacles, we can not only save lives, we can begin to imagine a new world. True innovation means that the discovery of new cures and remedies will take place not only in Stanford and Oxford but in Bangalore, Shanghai and Cape Town. It means the cities of the poor world will begin to reap the health and economic benefits of science. The greatest gift for poor nations is the scientific power to heal themselves.

CHAPTER ONE

Morogoro, a sleepy African city at the foot of Uluguru Mountains in central Tanzania, has little to distinguish it apart from one thing: it is one of the greatest breeding grounds for the most lethal mosquitoes on the planet. The tropical city of about two hundred thousand[1] people has the ideal warm, muggy weather for all kinds of mosquitoes, both those that annoy and those that kill. Swarms of these insects breed in the still waters of Morogoro's ponds and swamps, on the river Ngerengere and on the Mindu Dam, which was built by Western donors in 1979.[2] Thick clouds of them rise from the waters and bite the city's residents, usually in their homes after dusk.

One lethal genus, the *Anopheles* mosquito, is deadlier than all others and is common in Morogoro. To the untrained eye, it looks just the same as other mosquitoes, but the females carry in their saliva a tiny but potentially dangerous parasite. They awaken after dusk, quietly sneak into homes and gently alight on exposed skin, just long enough to prick the host, pick up a little meal of blood and leave behind the *Plasmodium* parasite, which quickly moves into the victim's liver and after about ten days causes malaria, a sickness accompanied by high fevers, chills and terrible flu-like symptoms. Most people who get malaria do not die; adults usually get better, especially if they receive treatment. But the female *Anopheles* carries the one type of parasite that is often lethal, the *Plasmodium falciparum*. It can be a merciless killer of children and of those already weakened by other illness.

The story of malarial mosquitoes is one that was very familiar to me, Abdallah Daar, long before I decided to leave Tanzania to pursue a career in medicine. I was born into a large, tight-knit family in Dar es Salaam, the tropical port and largest city in Tanzania, in eastern Africa. The fourth of twelve children of a butcher and cattle trader of Yemeni origin, I was the first member of my family to go to university. As a young boy, I remember waking up at four in the morning to help my father at the butcher shop, and then I'd go to school, sometimes with cattle blood still on my clothes. I was a dedicated student and I longed to discover the world beyond the confines of my small community. From the very beginning, my family elected me to tell their stories and to bring tales from the outside world back to our home. By age thirteen, I was devouring any reading material I could get my hands on, and I still recall with excitement the hallowed feeling of holding *Time* magazine in my hands and the anticipation that gripped me when I considered that one day I might find my place in the brave new world it depicted so far away.

My eldest sister, Alwiya, however, was to take a different path in life. When she was just sixteen, my father sent her to Morogoro, where she was married to a man a decade older. Later, she moved to another quiet, inland town, where she lived a traditional life in a dark earth-brick house, with young children jostling at her feet while she cooked curry and rice, or *ugali*, a cornmeal dough to dip in a thin peppery stew of dried fish or chicken. When I was a teenager, I'd visit my sister, and I am still haunted by the memories of our talks about malaria, which was a constant problem in her life. Her children had suffered various bouts, and though she couldn't afford expensive treatment, she counted herself lucky that her family could eat every day and that no one around her—so far—had died of the deadly forms of the disease, especially cerebral malaria, a common killer of children in sub-Saharan Africa.[3]

Throughout my teen years, I continued to study doggedly. My determination to be a scholar eventually paid off: I got into the best high school in the country, and after graduating, I attended medical school in Uganda. One day, Alwiya called me in Kampala to tell me that my eight-year-old niece, Fatma, was terribly sick. Fatma's heart was banging wildly in her chest, she explained. Alwiya pinned all her hopes on me, her brother the medical student. I arranged for my niece to be seen by a cardiologist, who diagnosed her with severe rheumatic heart disease; rheumatic fever had permanently damaged her heart valves. The disease was preventable; the valve damage could be repaired surgically. In fact, rheumatic fever and the damage it causes to heart valves is now rare in developed countries.[4] But this kind of health care was not available to someone like Alwiya, so despite my attempt to help, I was powerless to save my niece. A few months later, Fatma died.

I was horrified and saddened by this news, and as I continued my work at Mulago Hospital, the huge main teaching hospital of Makerere University in Kampala where I went to medical school, I began to see first hand the dire health concerns of Uganda's poorest people. Toddlers with swollen bellies and emaciated limbs lay listlessly in their mothers' arms, with flies buzzing around their open mouths as their mothers pleaded for help. These children suffered from kwashiorkor, a severe form of malnutrition. For four years, I attended to the illnesses brought on by hunger and malnutrition, but then Uganda's military dictator, Idi Amin, ordered all Asians out of the country. In the ensuing violence and chaos, I left, knowing that while I could escape the poverty and upheaval, millions of others could not.

I finished medical school at the University of London, and after a short spell teaching anatomy at the University of Texas, I went to Oxford for postgraduate training in surgery and internal medicine. After completing a doctorate in immunology, I followed up with a fellowship in organ transplantation. By 1979, I was a lecturer in Oxford's Department of Surgery. In the mid-1980s, I moved to the United Arab Emirates to start a biomedical research institute and a clinical organ transplant program. In addition, I helped to found the country's first medical school.

My career as a transplant surgeon was flourishing, and in many ways the dreams that I had had as a boy were turning into reality. In 1994, in Oman, I performed a kidney transplant, with two kidneys the size of beans, from the youngest and smallest donor in medical history, a baby born at thirty-four weeks gestation and declared brain dead soon after. The recipient of the kidneys, a seventeen-month-old boy from a poor family, is still alive and doing well sixteen years later. It was a remarkable surgical feat that required microscopic techniques, and this operation became an official world record. But even as I achieved professional success and a reputation as a surgical innovator, something in me was always drawn back to my memories of home, to the community and the family members I'd left behind. When I became homesick, I would speak to Alwiya on the phone and occasionally I even visited. By then, she had moved back to Morogoro—home of the lethal *Anopheles* mosquito.

As the only doctor in the family, though by no means an expert on malaria, I would advise my sister on basic prevention measures, like covering her children with bed nets at night, spraying DDT on the walls of her home and draining all nearby standing pools of water. But even as I recommended these measures, I couldn't help but feel there wasn't enough I could do—both as a doctor and as a brother—to protect my sister and her family.

Then, in December 1997, I was working in Oman, at Sultan Qaboos University Hospital, and living comfortably with my own family in a rented house in Muscat, the jewel of the Arabian Sea. My fifteen-year-old daughter, Nadia, was at home when the phone rang. Nadia struggled to understand what the caller was saying. It sounded to her like Kiswahili, but having been born in England and brought up in an English-speaking household, she had not yet mastered either Kiswahili, which I had grown up speaking at home, or Arabic, the native language in Oman. From what she could make out, it sounded as though her Aunt Alwiya had died.

It wasn't until I came home late in the evening that I learned the horrific news of my sister's premature death. It was a terrible shock, especially since I hadn't known in time to fly to Morogoro for the burial, which, according to Muslim tradition, took place within twenty-four hours of her death. But it was only when I called my brother-in-law in Morogoro that the grievously unnecessary cause of Alwiya's death became clear to me.

On a hot afternoon, Alwiya had complained of severe pain. "My body hurts, my bones hurt," she told her son. She was sure she had malaria, and as the fever began to drench her in sweat, my nephew took her in his minibus to the small outpatient clinic nearby.

The diagnosis was exactly as Alwiya had predicted, and she was given a bottle of antimalaria pills. Their active ingredient, chloroquine, was discovered in Germany in 1934,[5] and these pills continue to be the primary treatment option in resource-poor regions,[6] where they are both inexpensive and accessible. Alwiya took them as instructed, but she didn't get better.

After a couple of days, my family decided to take her to a nearby hospital, where doctors set up an intravenous glucose drip, the standard response to the decline in blood glucose that often accompanies malaria. What those doctors may not have realized was that Alwiya was diabetic; the glucose drip likely raised her blood sugar to a dangerous level. As well, the antimalaria drug she'd been given was useless in Morogoro; the local parasites had long ago developed resistance against the fifty-year-old drug.

What was clear to me as I listened to this chilling story was that a more sophisticated hospital would have been able to save Alwiya's life. But because she lived in Morogoro, Alwiya had died of a treatable disease less than four days after she'd first noticed symptoms.

The unexpected loss raised many questions for me: Could I have saved my sister had I been there by her side? Would she have lived if I'd arranged for her to be transported to a hospital in Dar es Salaam? As I thought about the possible outcomes and of the loss of someone so dear to me, the frustration mounted. Here I was, a world-class transplant surgeon who had tapped into the highest level of medical expertise to save a baby in Oman, but I felt powerless to save my own sister from a disease that wouldn't have killed her if she'd been living with me. I had entered the practice of medicine to save lives, and I helped hundreds of patients a week, improving the quality of their lives as well as prolonging their lives. But in the developing world, thousands of people died every week of the same illness that had just needlessly killed my eldest sister. Was I really using my professional skills to serve the people who most needed help?

My sister was just one of the estimated one million people who die of malaria each year, mostly in sub-Saharan Africa.[7] Her death was commonplace in a country that was still using an old remedy that the one-cell malarial parasites had long ago outwitted. Malaria was a "tropical disease," as the European textbooks bluntly put it, and that meant it didn't affect the rich world; therefore, in terms of research funding, it was largely neglected[8] at the time of Alwiya's death in 1997. Though about a quarter of a billion people around the world were afflicted each year, scientists, drug companies and affluent funding institutions spent little time or resources searching for new drugs or vaccines or better preventive measures beyond faulty mosquito nets and DDT.

What's worse, here was a disease that could actually teach science valuable lessons about the human genome, which just around this time was being mapped. Malaria forces the human genome to evolve mutations in hemoglobin, the oxygen-carrying molecule in blood, which partly protects against malaria.[9] By studying it, scientists could gain insight into the way that parasites learn to evade the human immune system. Surely the application of life sciences to malaria should have been a subject of intense scientific interest, and I was disheartened to find that it was not.

Around this time, I began attending my fair share of meetings at the World Health Organization's headquarters in Geneva, and I saw first hand how the research needs of the developing world were being ignored. Why did the scientific community believe that delivering substandard health service was enough and that research that could permanently end the spread of illnesses like malaria was not worth investing in?

What came to be known as "tropical diseases" had not always been ignored in the past—quite the opposite, in fact. A century ago, when Europeans were colonizing what is now the developing world, Europeans led the attack against malaria. A French army doctor, Charles Louis Alphonse Laveran, was the first to propose that parasites caused the illness, a discovery that earned him a Nobel Prize in 1907.[10] And just a few years before that, an English doctor working in Calcutta, Sir Ronald Ross, had proved that malaria was spread by mosquitoes, a finding that had also won him a Nobel Prize, in 1902.[11] His recommendations saved the lives of thousands of workers building the Panama Canal between 1905 and 1910.[12] Malaria and yellow fever, both transmitted by mosquitoes, were sickening thousands of workers on the canal and thus threatening the project. Through an integrated program of insect and malaria control, hospitalizations of workers due to malaria were cut from around 80 percent of the total workforce annually to around 10 percent. By 1910, yellow fever was eliminated among the canal workers and malaria was significantly reduced.[13]

Although disputed, it is generally believed that malaria was introduced to the Americas with the arrival of various explorers, colonists and slaves in the early sixteenth century, and by the mid to late seventeenth century it had also spread to British North America. The US Centers for Disease Control and Prevention (formerly known as the Communicable Disease Center) was formed in 1946 with a mandate to fight malaria, and it did, by draining swamps and spraying DDT on the walls of homes. Malaria lost the fight, at least in North America, and by the 1950s, it was mostly wiped out.[14] But while a model for control was clearly proved effective in North America, the global community did little to implement a similar model in Africa to curb the spread of the disease there.[15] So in Africa today, a child dies of malaria every thirty seconds.[16]

Malaria is just one of the Big Three infectious diseases of the poor that scientists had ignored, alongside tuberculosis and HIV/AIDS.[17] Tuberculosis was largely eliminated in the developed world by the end of the twentieth century,[18] but it continues to be a major scourge, killing about 1.7 million people a year, mostly in the developing world.[19] TB can affect many tissues in the body, including the brain and bones, but its most commonly lethal target is the lungs, which are ultimately destroyed by the infection. TB requires several months of treatment with a cocktail of toxic antibiotics.[20] The patients, who find it difficult to stick to the treatment regimens and often stop taking the medications, become lethargic, are feverish, have night sweats, lose weight, cough a lot, sometimes coughing up blood, and pass on the infection to others through droplets when they cough. While a huge proportion of people worldwide are infected, most people have the latent form and don't get sick.[21] The active form hits people in poor countries because their immune systems are weakened by poor nutrition, crowded conditions, general poor hygiene and concomitant infections like HIV.[22] Once people there get sick, accounting for the 9.4 million new tuberculosis cases each year (as documented by the WHO[23]), they can't obtain adequate treatment from under-resourced health systems.[24] And so the cycle continues, and more are infected.

Similarly, by the late 1990s, HIV/AIDS had become an increasingly manageable chronic disease in the rich world, thanks to powerful but expensive antiretroviral drugs. But in the poor world, it is still a massive killer. In the year Alwiya died, AIDS killed about one million people, most in sub-Saharan Africa.[25] In Botswana, one of the worst-hit nations, up to 30 percent of adults were infected. Without access to the life-saving drugs, their prospects for survival were poor. Life expectancy, even in this relatively wealthy African country, dropped from sixty-five years in the 1980s to thirty-five in the early 2000s[26]—compared with seventy-seven years in the United States.[27] According to the most recent UNAIDS data, there are about 33 million people living with HIV around the world, 2.7 million new infections each year and 2 million deaths.[28]

The list of deadly killers doesn't stop there. Millions in the poor world perish every year from a long list of so-called "neglected tropical diseases" with exotic names like onchocerciasis (river blindness), human African trypanosomiasis (sleeping sickness), soil-transmitted helminthiases (hookworm) and dengue fever, illnesses that few people in Western countries have ever heard of and are unlikely to ever contract. These illnesses affect *a billion* people every year.[29]

And of course, there's the duo of pneumonia and diarrhea that in rich countries rarely imperil young lives. Pneumonia and other acute respiratory infections kill an estimated 1.8 million children under age five each year,[30] most of whom live in the developing world. More than one half of these pneumonia deaths are caused by just two kinds of bacteria, pneumococcus and *Hemophilus influenzae*. Both can be prevented by existing vaccines and, when caught early in infected children, can be treated with antibiotics. Diarrheal diseases, caused by bacteria, viruses or parasites, are no longer a major problem in the developed world because of good sanitation, clean drinking water, healthy nutrition and education on hygiene measures such as hand-washing. Yet another 1.5 million children die every year of diarrheal diseases in the developing world.[31]

In the 1980s and 1990s, the world's scientific researchers were largely ignoring the poor, especially the 40 percent of humanity who were living on less than two dollars a day and were far sicker than the rich. In 1990, one study[32] revealed that only 10 percent of global spending on health research was used to study conditions in developing countries, even though people there suffered 90 percent of the global disease burden. This became known as the 10/90 gap. Citizens in the United States, Canada and Western Europe were getting the lion's share of health research dollars even though their people suffered only a small fraction of the world's diseases. In other words, if you were an impotent, depressed, rich adult man living in the West, science delivered for you, in the form of Viagra and Prozac and a host of other remedies. If you were a poor person in Tanzania, you likely died long before you had to worry about the onset of midlife health concerns.

One big problem I started to see was that drug companies in the late 1990s felt the world's poor weren't a viable market because they couldn't ever pay back the cost of discovering, testing and bringing to market new drugs, a burden that the major companies claimed cost them up to a billion dollars for each new drug. In the last quarter of the twentieth century, pharmaceutical companies developed 1,393 new drugs, but only 16 of them were created to combat tropical diseases and TB.[33]

Even when scientists discovered a remedy for a disease that affected both rich and poor, it took a long time to get it to the poor. When the hepatitis B vaccine was developed in the late 1960s,[34] for instance, it quickly became widely available in the rich world, and today is routinely given to schoolchildren. In the developing world, however, where most of the world's hepatitis resides and where the virus is a common cause of primary liver cancer, the vaccine is still not widely available. In other words, the people who need the vaccine the most are the least likely to get it.

I became determined to change this, and in 1998, a year after my sister's death, I left my high-powered post as chief of surgery in Oman to begin an uncharted journey into the world of global public health. My move coincided with the extraordinary acceleration of the genetic revolution that had been building for a century, ever since Gregor Mendel conducted his meticulous pea-growing experiments to show how traits—or genes, as we now know them—are inherited. I knew that genetic science was still largely confined to the most sophisticated labs in the rich world, but I also saw that this revolution could offer hope to developing nations.

In 1998, I began working with my colleague Jean-François Mattei on a report for the World Health Organization. Together, we explored the ethical, scientific, social and legal implications of medical genetics and biotechnology, and how these might apply to global health. By then the Human Genome Project was in full swing. Hundreds of scientists in the United States, Britain, Japan, France, Germany and later in China were collaborating to map out the entire sequence of the human genome. Many considered this monumental quest to crack the code of life a greater goal than splitting the atom or landing a man on the moon.

The mission was to convert the biological instructions inside the molecules of DNA into a form that could be read and understood. Genes govern inherited traits in humans, everything from eye colour to embryo development to vulnerability to breast cancer. Four nucleotides—adenine, thymine, guanine and cytosine—form the arms of the twisted ladder of the DNA double helix, but turning this minute architectural wonder into a readable language was a mammoth task. These nucleotides come in pairs, and there are a vast number of them—three billion base pairs are contained in the twenty-three pairs of chromosomes inside each human cell. Before anyone could read the four-letter genetic language effectively, scientists had to write down, in order, the genetic alphabet of the human genome. If I were to present that alphabet, the list of letters would amount to a stack of paperbacks two hundred feet high—or two hundred five-hundred-page telephone directories. If I recited the list at one letter per second for twenty-four hours a day, it would take me a century to finish.

Somewhere in that immense list of letters were the genes containing the coded instructions that govern all human life. Yet when the Human Genome Project[35] began in the early 1990s, scientists didn't have a powerful search engine to find them. Hunting for a gene was like looking for a specific sentence in a library—without the help of the card catalogue or Google. But by the late 1990s, scientists were creating powerful new sequencing machines and technologies to speed up the process.

The results were more than spectacular. It took roughly four years to sequence the first billion base pairs; the second billion base pairs were sequenced in less than four months. The digitalization of biological data meant scientists could now locate a gene and read it far more easily than before. Instead of hunting through the library stacks, you could just Google a phrase, and presto, the book title and page number appeared.

It was an exciting time, and the Human Genome Project was beginning to energize not only genomics but life sciences as a whole. It was spawning new fields like bioinformatics (the marriage of genomics and information technology), proteomics (the study of proteins in cells, tissues or organisms) and transcriptomics (the study of messenger molecules produced in a particular cell type).

The mission to sequence the genome had turned into a race between two groups of scientists from the rich world—an international public consortium led by Francis Collins, a geneticist who had helped to identify the genetic mutation that causes cystic fibrosis, and Celera Genomics, a private company led by the visionary thinker Craig Venter. When these scientists considered the possible applications of their awesome new abilities, they pointed to the potential to read the genetic profile of individuals and customize drugs for them. No one was seriously considering the implications for the 90 percent of humanity in the poor world who were shut out of the genome and life sciences game.

But as I watched all of this unfold, I couldn't help but see possibilities for the developing world. What if the discoveries made in the sophisticated genomic labs of the developed world were applied to the needs of ordinary Africans and other disadvantaged people? The fledgling sciences of bioinformatics, transcriptomics and proteomics could fundamentally alter our understanding of diseases and our ability to treat them. They would, I hoped, usher in a new era of better and less expensive drugs, vaccines and diagnostics—for the benefit of all people, no matter where they lived.

The old way of discovering drugs relied on trial and error, and a heaping measure of luck. Take the antimalaria drug chloroquine prescribed to my sister. The story of the drug began hundreds of years ago and the details have been lost to history. What is known is enmeshed in controversy. It is thought that the natives of Peru used the bark of the cinchona tree to treat fever. In the seventeenth century, likely through a network of Jesuits, samples found their way to Spain and Italy, where malaria, then known as marsh fever, had killed several popes and cardinals, along with a multitude of ordinary citizens. From then on, the remedy discovered by the Peruvians was renamed Jesuit's bark and used to treat malaria.[36]

The active ingredient in the bark, quinine, was isolated by French researchers in the early nineteenth century by applying solvents to the cinchona bark. By the early 1900s, organic chemists had figured out its crude chemical structure, but they weren't able to synthesize it to produce chloroquine until 1970. The full molecular structure wasn't worked out until 2001.[37] No matter what the starting point, drug development in the old days depended heavily on guesswork, followed by years of trial and error.

Today, it's a different story. In the case of malaria, for example, the advances of the Human Genome Project allow scientists to tackle the disease from various new angles. Researchers are studying how the genes of both the parasite and the human host are turned on and off as the parasite enters the body, progresses to the liver and then parasitizes red blood cells and, in deadly cases, affects the brain. They are studying genetic profiles to understand why some children get cerebral malaria and die while others don't. And most notably, they are attempting to alter the *Anopheles* mosquito's genes to make it incapable of transmitting the malarial parasite. These scientists are also using software to design vaccines based on knowledge of the parasite's genome. By using proteomics, researchers are working out the detailed structure of malaria proteins so that designer drugs can fit into all the nooks and crannies of the protein to neutralize it. And they are genetically engineering a synthetic version of the anti-malaria drug artemisinin that promises to be just as effective but less expensive than the original.

This new knowledge might make it possible to develop drugs at an affordable cost for the developing world. Yet in 1999, what Jean-François Mattei and I were forced to conclude in our report[38] for the World Health Organization was that "much of the developing world [was] excluded from the technology." Did it really have to be that way? I wondered. In the case of malaria, the basic materials, like the bark that contained quinine, originated in developing countries. The patients were there too. And now genomes could be read over the internet from anywhere in the world. If scientists in the developing world were trained in modern genomics, couldn't they too read the code and be part of the solution? Mattei and I stressed to the WHO and to the wider scientific community the importance of including developing countries in the genomics revolution to improve the health of humanity on a global scale. Despite our best efforts, the elite society of science paid little attention.

But to my great relief, there was another man who understood our cause. He and I would soon join forces on a mission to take genomic science on a brave new journey from lab to village.

My name is Peter Singer. My parents emigrated to Canada from Hungary during the 1956 revolution. Like many immigrants, they emphasized the importance of education and hard work, and were focused on creating a better life for me, their only child. My father was a dentist, and though he studied hard to pass the Canadian dental exams, no amount of preparation could help him get over the significant barriers constructed to keep immigrants like him away from lucrative professions. He had his first heart attack while preparing for one of these exams, and he died of a subsequent heart attack when I was just eleven years old. My mom was a dental technician and she made teeth, initially in our kitchen, from dawn to dusk. She saved enough to enroll me at an exclusive prep school. From the time I was a young man attending Upper Canada College in Toronto, I knew I wanted to be a doctor. It was there that I was first exposed to bioethics when my biology teacher assigned an essay on the ethics of research with human subjects. That was in 1978, before almost anyone had heard of bioethics.

After high school, I proceeded to medical school, where I trained in internal medicine and hoped to become a clinician and scientist.

I was planning to study the biophysical properties of the red blood cell. During my year as an intern, though, my understanding of what it meant to be a doctor began to change. I had always been interested in bioethics, but hadn't considered it as a career. Even before I graduated, however, I encountered many situations where advances in biology and medicine raised ethical controversies that the medical profession was simply not addressing fast enough.

I vividly remember one incident in particular. One of my patients, a young woman with widespread cervical cancer, was desperately ill and very concerned—not with getting better, which was impossible, but with her imminent death. Meanwhile, the doctors and nurses caring for her insisted on monitoring her blood counts and blood chemistries. No one dared discuss the issue she cared about the most—how she was going to die. When she was slipping away, the doctors and nurses decided not to resuscitate. I was shocked to realize that the medical practitioners around her, myself included, were called upon to make a decision that had never once been discussed with the patient herself or with her family. In a medically sophisticated world where technology was keeping people alive longer and offering a new array of choices to those with health problems, it seemed to me that my profession was lagging desperately behind in considering the ethics underpinning those choices. Along with many other medical practitioners, I joined a new movement afoot that aimed to advocate for patients' rights and include bioethics as part of quality care. I decided to switch my specialization to bioethics, and at the University of Chicago, I began training as one of the first doctors in medical ethics. I complemented this with training in public health and research methods at Yale.

In 1990, my training complete, I returned to Toronto, and by 1999 I had started the world-class University of Toronto Joint Centre for Bioethics, which was linked to ten large teaching hospitals in the city. It had an unusual approach—unusual at least for the field of bioethics: to work from inside the medical system to achieve practical change that would benefit real patients, especially those in their final days of life. It was at this time that I made a trip that would change forever my professional future.

While in Cape Town for South Africa's first national bioethics conference, I saw for the first time a clinic in an African slum. This was during the last throes of the dying apartheid regime. The clinic was a small room with peeling blue paint—and nothing in it. What, I asked myself, was the point of broaching the challenges of bioethics if South Africa still lacked basic health care for the poorest South Africans?

From that point on, whenever I attended conferences around the world, I made a point of visiting the slums. It was an eye-opener. In Bangladesh, I visited a hospital devoted to treating diarrhea. Known by locals as the "cholera hospital," it comprised one giant room filled with dozens of babies lying in cots, their mothers hunched beside them. Though the situation appeared desperate, this was actually a place of hope, a life-saving factory. For only ten dollars, each baby could be revived with oral rehydration. I was struck by the fact that in North America, millions of dollars were spent every day on extending people's lives, whereas in that Bangladeshi hospital, only a few hundred dollars a day was giving countless young children a chance to grow up.

The two parallel universes were hard to reconcile. A question surfaced in my mind: how can we, as a planet, justify that a rich person's final days in a city like Toronto are worth far more than the lives of thousands of children in a poor country like Bangladesh? I didn't have an answer, but I came to an uncomfortable conclusion:

I'd been spending much of my professional time on issues that were, on a world scale, irrelevant. The most immediate of all ethical challenges was the deep equity issue, the gap in health between rich and poor. Here was where I wanted to make a difference, if only I could find a practical and progressive way to do so.

In May 1999, I was sitting in my office in the former rectory of an old Toronto church where my bioethics centre was located. An acquaintance of mine, Bernard Dickens, a University of Toronto law professor and an expert on the ethics of organ transplantation, had set up a meeting between me and a hotshot surgeon named Abdallah Daar. I looked up to see Abdallah limp into my office on crutches. He immediately explained that he was recovering from reflex sympathetic dystrophy, a neurological disease of unknown origin that causes edema and swelling in the bone marrow. I rose to greet my guest, but because of recurring back troubles, I walked stiffly across the room. "You have a bad back," Abdallah observed. "And you have a bad leg," I responded. "We'll make a great team." This was the inauspicious beginning of an unusual but fruitful academic marriage.

As we got to know each other, it became clear that we shared a common goal: to bring the innovations of the genomics revolution to the developing world. At the bioethics centre, I had already begun a project that was bringing trained ethicists from India, Pakistan, Uganda, Nigeria and other countries to Canada to train. By this point, the centre had developed into a well-respected brand that was international in scope and constructive in its approach. Still, we had not yet broached the new territory of genomics. The Human Genome Project had set aside 5 percent of its annual budget to study the legal, social and ethical issues of genomic advances, but the projects that received funding always focused on issues such as informed consent, patents and privacy—the kinds of narrow legalistic issues neither

Abdallah nor I was interested in. After all, what did legal issues have to do with the billion people living on a dollar a day, people who lacked the most basic health care, clean water or even adequate food?

As we began to talk more seriously, Abdallah and I imagined the possibilities. What if we challenged our fellow scientists to consider the real issue: who would benefit from the profound changes that the genomic revolution offered? It would be a grand mission. We'd have to convince our scientific colleagues that the discoveries in the world's most sophisticated labs could be used for 90 percent of the world's population, not just for a few rich individuals. We'd also have to convince the scientific community that it ought to include scientists from the developing world in the discovery process. And most important, we'd have to stress one core value at the root of it all—that where you live would not determine how long you lived. It was an audacious goal, just the kind that we both relished.

To begin, we needed to find some research to launch our quest. We quickly found a study[39] published that year in *Science*, one of the world's leading scientific journals. Hassan Jomaa, a scientist living in Germany, had read part of the sequence of the genome of the *Plasmodium falciparum* parasite that causes the deadly form of malaria. He had discovered that one of the genetic instructions that coded for a specific enzyme was the target of a drug developed by a Japanese firm, Fujisawa Pharmaceutical Company, Ltd.[40] The drug they developed, fosmidomycin, had never been commercialized. Jomaa tested fosmidomycin on mice in his lab. It worked.

Here was an entirely new class of drug that had been discovered for a few hundred thousand dollars, compared with the billion dollars that a pharmaceutical firm might spend on such an enterprise. Early trial results on humans proved that fosmidomycin was effective and well tolerated in the treatment of patients with acute uncomplicated falciparum malaria. The drug was even more effective when combined with other drugs such as clindamycin.[41] We were thrilled by this because it meant the cost of conducting research could be cut substantially, which would make it possible to deliver drugs to the developing world at an affordable cost.

In March 2002, we learned of another important cost-saving measure through Dr. Eva Harris, an up-and-coming California scientist who was making great strides in proving that, with some ingenuity, scientists could find cheaper ways to bring innovations from high-tech labs to the poorest villages of the world. Harris, a professor of infectious diseases at the School of Public Health at the University of California, Berkeley, had won a lucrative MacArthur Fellowship, which many of us think of as "the genius grant." She applied her winnings to groundbreaking research that tackled the problem of how to make the polymerase chain reaction, or PCR—an essential technique in molecular biology that allows DNA to be amplified—more cost effective. Harris showed that instead of paying hundreds of dollars for the test materials to purify DNA, scientists could go to any hobby store and for five dollars buy a 20-pound bag of ceramic dust that would roughly accomplish the same task.[42] What was even more exhilarating was that Harris's low-cost version of the PCR technique had been used in Latin America to diagnose two neglected tropical diseases—leishmaniasis, a disease spread by sandflies that causes ugly and potentially fatal skin sores, and dengue fever, a mosquito-borne viral disease that causes potentially fatal hemorrhaging, fever, pain and flu-like symptoms. Harris's method allowed communities to track these diseases and take preventive action at an early stage.

Harris was a true pioneer, a shining example of how creative thinking can solve problems that have stymied science for years.

Here was irrefutable proof that modern biotechnology could in fact be deployed for the benefit of people in the developing world.

As we continued our search, we uncovered more signs of hope. Not only was research from cutting-edge labs being exported to developing nations but some research was actually happening there, too. The world's first vaccine against meningitis B had been created not in France or England or the United States, but in Cuba, at the Finlay Institute. The vaccine has since been exported and licensed for use in several Latin American countries.[43] And in India, a vaccine against a form of malaria common in that country was under development.[44]

As we surveyed all of these fascinating endeavours around us, we realized that we weren't alone—that other scientists and doctors around the world were awakening to the possibilities of improving global health. Pioneers like Eva Harris and Hassan Jomaa were building the first links between lab and village. It was time to convince the titans of genomics that they could do the same.

We began our campaign in the prestigious journal *Science* in 2001[45] by asking a provocative question: how do we harness genomics to improve global health equity? Shortly after, we hosted a meeting of geneticists associated with the WHO genetics program. The visiting scientists were unanimous in their negative opinion of our mission: in their view, we had veered way off track. Science progresses in the developing world, they said, because researchers focus on narrowly defined goals—dealing with one single-gene disease like sickle cell anemia, for example—rather than trying to harness the whole genomics revolution for the greater good. They reasoned that the world had seen other groundbreaking moments in the history of science and yet the health divide between rich and poor had not been resolved then. Did that not suggest, they posited, that thinking on a grander scale was futile? And finally, they asked, what did we know about the developing world? (To which Abdallah responded by waving his Tanzanian passport as proof of his origin.) To our absolute shock, we were being heavily criticized by the very scientists we'd assumed would support us.

Amid the skepticism, we followed up with a dare, one we playfully modelled after David Letterman's Top 10 lists, minus the satire. We conducted a study in which we asked the scientific community to rank the top ten biotechnological priorities for improving health in the developing world. (Of course, even suggesting that ranking such priorities might be a good starting point was a cause of controversy.) Our study was published[46] in 2002 in the highly regarded scientific journal *Nature Genetics*. Our priorities started with creating simple hand-held devices to conduct quick, low-cost checks for infectious diseases, and using simple technologies like the one Harris had used to diagnose leishmaniasis. Our wish list continued with a call to genetically engineer vaccines to stop big killers like malaria, HIV and tuberculosis, and also to develop vaccines that could be delivered without needles, refrigeration or multiple injections. We called for research into edible vaccines—ones that could be incorporated into potatoes or other staple foods to protect populations against hepatitis B, cholera, measles and other ailments. We asked for vaccines and vaginal microbicides that would allow women to protect themselves from sexually transmitted infections, and for computer-based tools to mine data on human and non-human gene sequences for clues on preventing and treating infectious and non-communicable diseases. To combat environmental contamination, we requested research into genetically modified bacteria and plants that would filter contaminated air. Finally, we underlined the need for genetically modified staple foods—such as rice, potatoes, corn and cassava—with enhanced nutritional value that would improve the overall health of millions of people worldwide. It was a bold article, and one that invited controversy by emphasizing not the grand accomplishments of Big Science but its colossal shortcomings. But the scientific establishment, still awakening to the huge potential of genomics, simply declared our list utopian. We were more convinced than ever that it was practical and achievable.

In 2000, the whole world listened with rapt attention as President Bill Clinton marked the significance of the draft sequencing of the genome during a White House speech[47] in Washington, D.C.: "Today we are learning the language in which God created life. We are gaining ever more awe for the complexity, the beauty, the wonder of God's most divine and sacred gift. With this profound new knowledge, humankind is on the verge of gaining immense new power to heal." It was a historic occasion, eloquently summarized, a moment of sheer, unadulterated scientific triumphalism. But while ordinary citizens were inspired by this momentous feat, there remained a reluctance among the scientific community to apply this new power to the biggest health challenges the world faced. Clinton had not addressed the real difficulty of translating the ability to read the genome into developing actual drugs, vaccines and diagnostic tools. And, most notably, he hadn't said a single word about using this "immense new power to heal" to help the people of the developing world.

But things were beginning to improve a little in enlightened circles. Francis Collins, one of the main geneticists credited with sequencing the human genome, and his colleagues, published a paper in April 2003 in *Nature* entitled "A Vision for the Future of Genomics Research."[48] Collins had held discussions with about six hundred scientific and political leaders from government, academia, non-profit organizations and the private sector to get them thinking about what to do with the power newly uncovered by the Human Genome Project. They posited that, to start, scientists needed to understand how the language of genes functioned. Next, they needed to translate "genome-based knowledge into health benefits." Collins and his co-writers went on to challenge the research community to understand the role of genetic factors (and non-genetic ones) in human health and disease. Understanding this would lead to the development of diagnostic methods to predict a person's susceptibility to disease and his or her response to a particular drug.

All these suggestions were interesting, but the finding that really caught our attention was this: the authors expressed the important goal of developing "genome-based tools that improve the health of all." They continued by saying that "incorporating this information into preventive and/or public-health strategies would be beneficial," and "understanding the genetic factors that make people susceptible to a disease like malaria could help scientists develop a drug to treat it and have a significant global impact."

We were also intrigued by a drawing included in the paper. It featured a house that looked like a modern research institution on a lush green campus. The pillars and floors of the house represented modern research, suggesting citizens could take shelter within this new house of science. But from the point of view of the developing world, the depiction might as well have been of an exclusive mansion, a Beverly Hills vacation home for the rich. Nowhere, it seemed to us, was there a room for the poor.

We continued to worry that if nothing was done about it, the new genomic language and the life sciences revolution would widen the knowledge divide and the related gap in health between rich and poor. The rich would get designer drugs, tailored for their personal genetic makeup; they would get longevity and a freedom from health complaints that strike at middle age. The poor would continue to die prematurely of easily preventable diseases. Science might, once again, fail the developing world.

We were now galvanized in our goal to build a better house, one that would include plenty of rooms for the 90 percent of the world's population that lives in poor countries. In 2003, soon after Collins' article in *Nature* was published, we started the Canadian Program on Genomics and Global Health within the University of Toronto Joint Centre for Bioethics; later, the McLaughlin-Rotman Centre for Global Health, also in Toronto, served as our base for this project. (And, gratifyingly, Francis Collins in 2009 went on to become the Director of the US National Institutes of Health, where one of his top five priorities is global health.) As researchers, teachers and global citizens, we hoped to publicize the failing health of the poor and the developed world's failure to help them help themselves. We wanted to bring to light some of the amazing research being conducted in the most surprising places, research that right now is offering real hope for millions of suffering people. We wanted to understand what the roadblocks were in both the labs and the villages, and how to get around them. But before we embarked, we needed a guide, someone focusing on the big picture, someone who might point us in the right direction.

Calestous Juma knows more than anyone else on the planet about transferring knowledge from lab to village. This handsome, charismatic, energetic, confident man has become a highly influential scientist and technology guru as well as an adviser to the secretary general of the UN, the U.S. Department of State and many African governments. One of the few Africans ever elected to the Royal Society, the 350-year-old academy of sciences whose members have included Isaac Newton and

Charles Darwin, he became the executive director of the Secretariat of the Convention on Biological Diversity, based in Montreal, before moving to Harvard as professor of international development. He also happens to come from a traditional fishing village in Boonyana, Kenya, not far from President Barack Obama's ancestral home.

When Juma first visited us in Toronto in 2001, we immediately knew he could help us. He quickly explained a significant problem we'd never quite considered before. He pointed out that there are innovators—those who generate scientific solutions to problems—and there are recipients—those who use the solutions to effect change. Science tends to focus on the creative ability of innovators while ignoring the fact that recipients of technological solutions must be similarly innovative to effectively implement a new tool in their milieu. Juma put it to us this way: "If I'm used to waking up in the morning and then just walking off to my farm, but suddenly there's a new technology that requires me to brush my teeth in the morning, that's a change in my lifestyle. It's an adjustment I have to make, to internalize that activity. I think our models tend to ignore that the user has to be equally as innovative as the inventor. Innovation should take place on both sides, not only on the side of the technology but on the user's side as well."

What Juma wisely notes is that scientific discovery can solve only part of the problem of how to move ideas from the lab to the village—it accounts for maybe a quarter of the resolution. To turn innovation into practice on the ground, communities have to be engaged from the outset. Yet rich countries typically ignore the ability of the recipient to absorb the great idea. The distribution of vaccines is a perfect example. Vaccines often don't reach the target village simply because there's no infrastructure—a lack of not only roads but also know-how. As Juma cautioned, "Dropping in a technology isn't going to help if the absorptive capacity isn't there."

These were words that echoed again and again as we embarked upon the journey we describe throughout this book. We discovered that science and technology would play only one part in offering solutions for health and development, whether through the use of the Internet, the introduction of cellphones or the discovery of genes. Once the science was in place, we'd soon encounter other roadblocks. How, for instance, could we be sensitive to the cultural values of individual communities and encourage village elders to support scientific innovation in their communities? How could we overcome the deep-rooted suspicion of multinational companies that fund the development of drugs, vaccines and diagnostic devices? How could we persuade companies in the developing world to embrace pioneering techniques to cut the cost of vaccines and drugs and so make them affordable? And how could we deal with the political obstacles blocking the road from lab to village—corruption, competing political priorities, inadequate public infrastructure for health care, and anti-biotech lobbies that so often stand in the way of progress?

We believe more than ever that these obstacles can be overcome. But ultimately, the solutions must come from scientists and communities in the developing world—the lab needs to be part of the village. As Juma suggested, we need to prepare communities for the change, and the change could be profound. We are no longer talking about improving bed nets or draining swamps near residential areas to ward off malaria; we are talking about tweaking the genome of the *Anopheles* mosquito to stop it from transmitting the disease in the first place. We're talking about fundamental and permanent solutions to diseases that have plagued the developing world for centuries. Changing the DNA of the ubiquitous mosquito involves profound scientific and ethical issues, as we will see later in this book. If these ethical, cultural and political issues can be resolved, then science can save millions of lives.

When we started this book, we thought we'd have to convince people in the rich world that they should help bridge the health gap out of a sense of global solidarity. As we show in the next chapter, generous benefactors such as Bill and Melinda Gates are often altruistically motivated. But, as Juma reminded us, people in the rich world should also act on the basis of their own "enlightened self-interest." In other words, what happens in the poor world will ultimately affect the rich world. The H1N1 flu bug, for example, didn't require a passport. It flew around the world in a flash and became a prime example of the globalization of infectious disease.

As we see it, health can ultimately become a powerful tool of development. But because people in poor countries have been shut out of the search for a cure or a remedy, they are shut out of the scientific industry that could develop one key part of the local economy. They don't have a chance to translate their ideas into products, companies and jobs for their communities. Without direct participation at the ground level, countries stay poor and dependent on rich countries.

In a prescient statement[49] more than thirty years ago, Nobel Prize–winner and Canadian prime minister Lester Pearson said, "There can be no peace, no security, nothing but ultimate disaster, when a few rich countries with a small minority of the world's people alone have access to the brave, and frightening, new world of technology, science, and of high material living standards, while the large majority live in deprivation and want, cut off from opportunities of full economic development; but with expectations and aspirations aroused far beyond the hope of realizing them." Developing countries have supplied the raw materials and even the human subjects for some of the world's greatest scientific advances, such as vincristine,[50] a drug used routinely to treat cancer, but in the business of scientific discovery, those subjects have been shut out for too long.

For us, this has become both a personal journey and a deeply moral quest, and while we are scientists, we often return to spiritual sources for direction. The great Reverend Martin Luther King reminds us that "all life is interrelated. . . . Whatever affects one directly, affects all indirectly." Perhaps Dr. King knew he was echoing the African-centred world view known as Ubuntu. It is characterized by a sense of self that is collective; a clear sense of one's spiritual connection to the universe; a sense of mutual responsibility; and a conscious under-standing that human abnormality is any act that is in opposition to oneself and one's fellow humans, or in opposition to God. It is this affirming view of the human community that has often kept us going in our mission to bring the lab to the village. The message is enshrined in the maxim "Umuntu ngumuntu ngabantu"—a person is a person through other persons, or, simply put, "I am because we are."

CHAPTER TWO

Towards the end of 2002, an intriguing invitation landed in our inbox. It was from Dr. Richard Klausner, the head of the Bill & Melinda Gates Foundation's Global Health Program. Klausner, a former director of the National Cancer Institute, was one of the most powerful and influential men in science, and he wanted Abdallah and me to fly to Washington, D.C., to meet him. He didn't say why.

Two weeks later, we met Klausner in a Washington hotel lobby. At fifty-one, he still had the brilliant mind of a hyperactive teenager. He explained that Bill Gates, the CEO of Microsoft, was planning a bold new strategy for the Bill & Melinda Gates Foundation. Gates had been spending hundreds of millions of dollars on the development and delivery of vaccines and other health technologies in the poor world, and would still do so. But now he wanted to take an even more ambitious step, to fund discovery science that would benefit the developing world—the science that could potentially lead to new vaccines, drugs and diagnostics for the diseases of the poor. We were amazed. Now here was truly thrilling news, we thought: no one had ever done this on a large scale before.

Klausner soon cleared up the mystery of why he had invited us here. The big question now, he said, was how to hand out the Gates Foundation money. Gates did not want to fund scientific research in the traditional way, as the U.S. National Institutes of Health would do, by supporting open-ended research projects driven by scientific curiosity. He wanted a more focused, goal-driven approach—a little like President John F. Kennedy's 1961 challenge to land a man on the moon before the end of the decade. Perhaps this would induce scientists to use the powerful tools they had acquired through the Human Genome Project to save millions of lives in the developing world.

There was a famous precedent for this approach, and it came from the world of mathematics. Hilbert's problems,[1] a set of ten mathematical problems (later expanded to twenty-three), were unveiled by German mathematician David Hilbert at a mathematics conference held at the Sorbonne in the summer of 1900. Over the next century, mathematicians around the world would try to solve Hilbert's problems, and they succeeded in all but two cases. (Four were considered too vague to solve.)

Gates, Klausner told us, loved the idea of a grand challenge. If Hilbert's problems worked so well to focus mathematical minds for a century, why not challenge the world's scientists to solve the biggest problems in global health? The Gates Foundation would determine a set of Grand Challenges for Global Health, to galvanize the world's scientists to attack the root of problems such as malaria and tuberculosis and infectious diseases like the rotavirus, the killer of so many children. But how would they choose the problems?

That was why Klausner had invited us to this hotel. "We need to have a way to identify grand challenges," he told us. "Your biotechnologies study seemed like a very good start." Klausner had just read our "Top 10" study,[2] published earlier that year in *Nature Genetics*. Klausner wanted to know how we'd selected our top ten. We explained that we had used a social sciences technique called the Delphi method,[3] which was invented at the beginning of the Cold War to forecast the effect of technology on warfare. It's a structured way of building consensus among a group of experts, one that avoids the interpersonal dynamics of meetings. For that study, we had chosen thirty-nine panellists from the developing world—experts in genomics, biotechnology and health—and asked them to identify and rank the biotechnologies that would most improve global health in the coming decade. About a hundred ideas came in from around the world. Then we asked the panellists to select the top twenty and provide any helpful comments on their choices. After looking through this list, we narrowed it to twelve. We asked the group whether they wanted to change the ranking, and then we settled on the final ten.

Our Top 10 biotechnologies were practical innovations that could make an enormous difference in the developing world in just a few years, we thought—if only the scientific world would pay attention. Our selection process had yielded a useful and eminently practical list. Could it help the Gates Foundation devise an entirely new way to fund research in global health? We didn't know the answer that day, but after talking with Klausner, it was clear that Gates was on the verge of launching something big and historic in global health, and we wanted to be part of it.

Gates acted quickly. Only a couple of months later, in January 2003 at Davos, Switzerland, the ski resort where the world's most powerful men and women gathered for their annual convention, Gates announced that his foundation was pouring $200 million[4] into the Grand Challenges in Global Health. This was the greatest single investment in history in the scientific discovery of solutions to diseases plaguing global health. "This initiative is about discovery and invention," Gates said on this thrilling day. "It is about finding specific solutions to the hardest problems. By accelerating research to overcome scientific obstacles in AIDS, malaria and other diseases, millions of lives could be saved."[5]

Soon after, just as we had hoped, we were asked to help. The Gates Foundation needed to determine what the grand challenges would be. We knew this would be a significant task, because the choice of the grand challenges would guide the next crucial decisions: what kinds of research projects would be funded, and who would do them. There were many unresolved questions. Upon what basis would the grand challenges be chosen? Who would propose and choose them? And would scientists from the developing world have a say?

Two months later, Rick Klausner invited Peter to an unmarked office in a modern high-rise in downtown Washington. He was nervous; he'd be presenting our ideas to three of the most powerful and influential men in science, and describing how the method we had used to identify our Top 10 biotechnologies could be used to decide upon the Gates Grand Challenges. As Peter stopped at the building's salad and sandwich bar, he met one of the scientists who would be helping to make those decisions.

Harold Varmus is one of the most impressive figures in modern science. He shared a Nobel Prize in Medicine in 1989[6] for discovering the cellular origin of retroviral oncogenes, genes that can cause cancer. Then, as director of the U.S. National Institutes of Health, which controls over half of the total global investment into health research, he doubled the NIH budget. After that, he became CEO of Memorial Sloan-Kettering Cancer Center in New York City. He is now the Director of the National Cancer Institute. Varmus, a tall, thin scientist who sometimes wears bike shorts to meetings, also had a subversive side. For example, he didn't see why research paid for largely by government should end up only in journals to which scientists and university libraries had to pay exorbitant subscription fees. As chair of a new online journal called PLoS (for Public Library of Science), he helped to create open-access publishing, enabling any reader to see a research paper for free.

After lunch, Peter moved upstairs to meet the other two men who would be joining the group that afternoon, Klausner and Elias Zerhouni. Zerhouni was born in a mountain town in western Algeria and fell for science as a boy after his uncle, a radiologist, showed him some early computed tomography (CT) scans. He eventually became a well-published radiologist in the United States and then vice-dean for research at Johns Hopkins University, before President George W. Bush appointed him to succeed Varmus as head of the NIH. In 2010 he was appointed U.S. Science Envoy by the Obama administration. In January 2011, he was appointed to President, Global Research and Development at Sanofi-aventis. An imposing, senatorial figure, Zerhouni has impressive political acumen. He also has, as Peter would soon find, an uncanny ability to see things outside the field of vision of ordinary scientists.

In the small library, Peter set aside his frustration over some last-minute technical problems with PowerPoint projection and handed out paper copies of our Top 10 biotechnologies to kick off the discussion. The key issue now was how to define a grand challenge.

They couldn't just be invented, as Hilbert had done a century earlier. There needed to be a logical way to decide on the problems. As Klausner, Varmus and Zerhouni thought through the issue, they ruled out certain things. A "challenge" couldn't be blue-sky research, the kind that is routinely funded by the NIH. It couldn't be a small technical problem that could be solved quickly. And it couldn't be a single big goal like Kennedy's goal of landing on the moon.

After a couple of hours, Varmus pulled out an idea. What about focusing on the missing piece—the one missing piece that's blocking the way from where we are now to where we want to be? He called such a piece a "critical barrier"—the roadblock or bottleneck. To identify the critical barrier, you needed to know enough about what you didn't know. This critical barrier needed to be big enough to challenge the planet's greatest scientific minds. It could not be specific to one disease, either.

This was the breakthrough we'd been searching for. The critical barrier ended up being the litmus test for the research that flowed out of the Gates Foundation's $200-million gift (later expanded to $450 million, after Bill Gates doubled its funding and British and Canadian sources contributed more money). The Gates Foundation would ultimately define a grand challenge as a "specific scientific or technological innovation that would remove a critical barrier to solving an important health problem in the developing world, with a high likelihood of global impact and feasibility."[7]

Now we turned to the question of who would decide what the grand challenges would be, and how. As we considered the "how," we agreed that the method behind the Top 10 was a good start, but Varmus wanted to modify it. He thought the Delphi method was too rigid for our purpose, and might not allow the deciding scientists any genuine opportunity to ask the kinds of questions that could influence the process. And again we circled back to who those scientists would be. Peter ran through the list of men and women from the developing world who had participated in our Top 10 study. The three senior scientists listened carefully; they had never heard most of these names, but they were eager to tap into this pool of scientific creativity—with one proviso: they themselves wanted to play a bigger, more influential role in defining the grand challenges than Abdallah and I had done in defining the Top 10.

Finally, Zerhouni, who had been listening quietly throughout the first two hours of the meeting, asked the classic question that is often posed in meetings like this: "Peter, if you had your Top 10 study to do all over, would you do anything differently?"

Peter immediately thought of the striking contrast between our study, which we had conducted on a shoestring, and the impending Gates Grand Challenges, funded by the richest man on the planet, with help from these three scientific titans. "Yes," Peter replied with a smile, "if I had it to do over, I would put $200 million behind our list." Everyone laughed.

After that Washington meeting, we knew two key things: Every grand challenge would have to deal with a critical barrier to qualify; and the grand challenges scientific advisory board would be made up of scientists in both the developing and the developed worlds. Many of the scientists from the developing world were chosen from the list of experts we had approached for our Top 10 biotechnologies. Now we needed to take the next step and ask the world's scientists to propose what the grand challenges should be.

We were asked to manage the process. In May, the Gates Foundation called for ideas, using journals, listservs and personal contacts. These ideas would serve as the basis for identifying the grand challenges. The response was immense: more than a thousand suggestions flowed in from around the globe. With the help of Tara Acharya, an Indian-born molecular biologist from Yale University who was working with us in Toronto, we analyzed the submissions. Some aimed to strengthen the scientific resources of developing nations. (Varmus ruled those out: "We're here to solve problems, not to build capacity.") We rejected others because they focused on a specific disease, like HIV; grand challenges needed to address a bottleneck issue that could affect many diseases, not just one. Other proposals seemed, at first glance, to be compelling. They tackled neglected diseases, the long list of ailments that had attracted very little research money to date. We were eager to include these on the list of contenders, since the impoverishment of neglected diseases was one of the reasons we had begun this journey. Yet we couldn't identify a critical barrier, and so neglected diseases as a group didn't make the cut. Proposals to curb the rise of smoking in the developing world met with the same fate, for the same reason. We even had to say no to the grandest grand challenge, from our point of view—the delivery of health to people in the villages of developing countries. Without a successful delivery system, this entire exercise would do nothing to help the poor. Yet the scientists who had crafted those delivery-oriented grand challenges couldn't come up with a viable critical barrier. Their proposals didn't make the cut, either.

The sheer number of ideas made it abundantly clear that Bill Gates had captured the attention of scientists around the world. The $450 million in funding meant that some of the most exciting scientists in the world were now thinking seriously about how to use great scientific advances in the lab to address the health disasters in the developing world. Many of their ideas dealt with the staple of global health, vaccines—how to make them easier, cheaper and safer to deliver, and how to make new vaccines for the diseases of the poor, such as malaria, HIV and TB. Yet many of the scientists were clearly struggling to deal with the concept of the critical barrier. It was a new and unfamiliar way to think about a scientific project.

As we analyzed more and more proposals, we came to realize that, to succeed, a grand challenge also had to make sense in the world where it would be delivered. Take, for example, the proposal for a male contraceptive pill. This idea was interesting enough to be assigned to Peter, as a member of the scientific board, for analysis. And for this task, Peter was lucky to have an extraordinary partner, one of the distinguished members of the scientific board, Bill Foege. The six-foot-seven American epidemiologist, who had directed the Centers for Disease Control from 1977 to 1983, is a global health hero for leading the campaign to wipe the scourge of smallpox off the map. Now Peter was helping him decide whether the male pill would be a worthy grand challenge. He sent Foege an email with some thoughts: "Male contraception may seem like a good idea biologically but the real social challenge is empowerment of women," Peter wrote. "There is a male contraceptive in existence, and it also protects against sexually transmitted disease. It is called a condom. The problem is that it is controlled by men in a patriarchal social structure. From a social point of view, vaginal microbicides look a lot better to me because they are controlled by women and also protect against sexually transmitted disease."

Foege responded to these views on the patriarchal social structure with a succinct little joke: "Thanks, Peter. I have always thought we should put estrogen in the water supply!"

The aim was to have a short list ready for a first, three-day meeting of the twenty-nine international scientists on the scientific advisory board in August 2003. Even as we counted down the days, Harold

Varmus continued to edit the list of grand challenges that would be submitted to the board.

Finally, on August 16, the two of us, along with an elite group of scientists from both rich and poor countries, drove to the Airlie Center in Virginia, a conference centre set inside 2,500 acres of lovely woods and gardens that was once described, in *Life* magazine, as an "island of thought."[8] The lineup of scientists on the advisory board was extraordinary. It included some of the most prominent figures in the developing world, such as Dr. Nirmal Ganguly, then director of the Indian Council of Medical Research, and Dr. Yongyuth Yuthavong, who would later become Thailand's science minister, along with American scientific stars such as Dr. Anthony Fauci, who has led the scientific fight against AIDS and advised President George W. Bush on the 2001 anthrax attacks and their aftermath. As the scientists sat down in the conference room, Varmus announced that we would have a difficult, historic and inspiring goal to reach in the next three days. We would have to select grand challenges from the list we had detailed in the thick black binders in front of each seat at the conference table.

We already knew that a significant number of the grand challenges would deal with vaccines. Moreover, Bill and Melinda Gates had clearly signalled their interest in vaccines by investing more than $1 billion in the GAVI Alliance (formerly the Global Alliance for Vaccines and Immunisation), which aims to accelerate the uptake of new and underused vaccines and to strengthen the ability of health care systems to deliver them. Now it was up to Sir Gus Nossal, the great Australian immunologist who had helped to launch the GAVI Alliance, to make the case for continued investment in this area. Before long, he had convinced the group that nearly half of the grand challenges should be devoted to vaccines.

The vaccine issue can be split into two related goals, Nossal explained. The first goal is to improve the delivery of existing childhood vaccines. Even on a poor continent, most people manage to get their children vaccinated. Yet they do it at great cost. Families sometimes have to walk many kilometres, three or more times, for each vaccination. It also costs the health system a fortune because the vaccines have to be refrigerated along the way. You could buy three to five times as much vaccine for the same amount of money if you didn't need to pay the cost of refrigeration. What's more, needles create a major waste-disposal problem, and if they're reused to save money, they also cause an increase in such diseases as HIV and hepatitis B.

Modern science has the ability to solve these problems, Nossal proposed. Advances in materials science and the preparation of biological materials can potentially stabilize vaccines so that they don't need to be refrigerated en route. Needle-free vaccines, such as nasal sprays, can be developed because of multiple advances in drug delivery as well as new ideas for activating immune cells in skin.

Nossal also proposed that we might take on a far bigger challenge: to give a vaccine in a single dose soon after birth, rather than with successive booster shots. Although scientists today have a deeper understanding than ever before about how antigens, which are foreign substances, stimulate the production of antibodies, the early spotters in the immune system's artillery forces, for many diseases we still do not know how to inoculate someone effectively with a single dose of an antigen. One obstacle is that we don't know enough about the workings of adjuvants, the mysterious substances—often held in a proprietary manner by vaccine companies—that strengthen the immune response.

Nossal was eloquent and persuasive, and he didn't have any trouble selling the group on the merits of the first three grand challenges—making vaccines needle-free, refrigeration-free and deliverable in a single dose soon after birth. "It was really a no-brainer," he said afterwards. "These challenges should be there."

But he wasn't finished. Nossal now turned to the second goal: developing *new* vaccines for diseases of the poor. Scientists have developed vaccines for many of the most common diseases in wealthy countries, he noted, but not for the world's biggest killers, which target people in poor countries. As of 2003, there were still no effective vaccines on the market for malaria, TB and HIV, which together kill about five million people every year.

But what is the best way to create new vaccines to tackle these immense problems? There were three promising lines of inquiry among the proposals we'd received. The first was to create new antigens—substances on the surface of toxins, bacteria and foreign blood cells that induce an immune response. The molecular revolution in biology, and especially the ability to read the sequence of genes in an organism that causes disease, has given scientists tremendous power to investigate this idea. Now they can identify the DNA sequence of the genes that code for the protein antigens and prompt the immune response, and visualize its structure. Yet the usefulness of this knowledge depends on being able to manufacture an antigen that delivers a *predictable* response from the immune system, something that has thus far eluded researchers.

Another line of inquiry looked at how you test a vaccine once you've developed one. Scientists can now create live vaccines in the lab that can theoretically stimulate an immune response without—we believe—making the person sick. But scientists have no way to test them for their safety and efficacy in humans, short of trying them out on human subjects. The challenge, then, is to create a model system, in vitro or in a mouse, that could test these live vaccines to see whether they are safe and effective. This isn't easy. Although drugs and vaccines are tested on mice before they are tested on humans, these mice have mouse immune systems. If, on the other hand, mice were genetically engineered to have human immune systems, they would be much more likely to demonstrate whether a vaccine would be effective in humans. This could speed up vaccine development in global health.

The final line of inquiry related to vaccines was the inverse of the others: it was the challenge to develop a test to tell whether a person is immune to a disease.

These important and substantial grand challenges on vaccines were familiar to anyone who had been involved in global health. The next idea, though, was a mind-bender for many of us—and yet it was just what we had had in mind for a grand challenge. It came from Dr. Fotis C. Kafatos, an influential Greek molecular biologist who was then directing the European Molecular Biology Laboratory, one of the world's highest-performing research and training labs. Dr. Kafatos has dedicated most of his professional inquiry to the study of malaria, in particular to the distribution system of the disease—the ubiquitous mosquito. Diseases spread by insects, he began, are exceptionally hard to eradicate, or even control. The World Health Organization has tried to stop the mosquito from spreading malaria, mainly through the use of insecticides. "Yet the eradication campaign has failed miserably," Kafatos said. Mosquitoes still spread diseases that kill millions of people every year.

"There's no magic bullet," he continued. No single strategy will work against disease like malaria or dengue fever. We need an integrated approach, a multi-pronged strategy, and, he said, it should include a radical new approach—changing the genome of the mosquito: "There's no way disease can be controlled and eradicated without addressing the vectors."

This isn't the first time that scientists have turned their attention to the insects that spread disease, Kafatos noted. The New World screwworm fly, a parasitic fly that eats the living flesh of warm-blooded animals, once ravaged livestock in the United States, but it was eradicated from American soil in 1966 by a program to irradiate male flies to sterilize them.[9] "It became very clear from this success that getting rid of the vectors was really important," explained Kafatos.

Today, he proposed, genetic manipulation had the potential to be just as effective. Laboratory experiments have demonstrated that genetically modifying the mosquito can substantially reduce its ability to spread disease. Still, the scientific challenge of creating a mosquito unable to transmit malaria and other diseases was immense. To begin with, genetic manipulation tends to weaken the insect. How do you make sure the genetically modified mosquitoes can take over the mosquito population? And what trait, exactly, would ensure that future generations of the insect were incapable of spreading a disease like malaria?

Kafatos knew he had to identify a critical barrier, and he had come with a prepared text to describe it: "We have not solved the full range of problems that would allow us to either replace an insect vector population in the field with one incapable of transmitting a disease aspect, or to control insect vector population numbers by genetic approaches. We also cannot accurately predict all of the ecological consequences of replacement."

This, we thought, was indeed a grand challenge. It was just the kind of innovation we'd been hoping to see, and it was challenging in every respect, not just scientifically. This would be a profound ethical challenge too: how would we make sure such a modification was done safely, with the full and informed consent of the community?

Most of the scientists in the room were comfortable with the idea of genetic modification, yet Kafatos sensed that there were some lingering doubts. "All of us were sensitive to the possibility that this proposal could generate problems politically," he said later. Yet "the most important disease, malaria, is such a terrible burden on humanity that we had to use ideas that may seem beyond the pale." Soon, Kafatos had convinced the group. The vectors, or insects, that spread disease would be attacked either by genetic manipulation or by a chemical strategy.

But the next proposal proved even more controversial. The Kenyan crop scientist Florence Wambugu tabled a plan to genetically modify crops. There was no doubt that this was a politically loaded idea. Monsanto, the global agribusiness, had just whipped up a storm of protest in Europe over its genetically modified crops. While the scientists in the room were not concerned about GM techniques, they knew that altering the DNA of food on the table—or insects that transmit disease, for that matter—could be politically controversial. Wambugu, though, is no stranger to controversy. She is a larger-than-life advocate of genetic modification of crops in Africa, and before this meeting she had handled her share of opposition, especially from European GM critics. But now, facing some of the world's most successful and powerful scientists, she would have to dig deep to make her case.

Wambugu's case was, of course, compelling. The poor in developing countries often eat a single staple food, like cassava, sorghum, bananas or rice. They don't have a rich diversity of foods on their plates, as most North Americans and Europeans do. These staples usually lack important micronutrients such as zinc and iron, vitamins such as vitamin A, and protein. The deficiencies cause many child deaths, weaken resistance to childhood infections and have long-lasting effects on children's intellectual development, leading to lower work productivity later in life and locking poor countries into poverty. Genetic engineering could change this picture, Wambugu told her audience. The staples of life could be tweaked to add the micronutrients and vitamins that the poor so often lack. "I was clearly encouraging them to think out of the box," Wambugu said later. "I was among the few at the meeting who saw food and nutrition as having a real and direct link with health."

Many of the scientists in the room were skeptical. Some—in a reflection of how the scientific establishment had divided itself into isolated groups, or "silos"—did not see a direct link between nutrition and health. Others questioned whether genetic modification would ever be widely acceptable in Africa. A third group questioned whether this initiative qualified as a grand challenge. After all, hadn't this already been done before? Two scientists, a Swiss and a German (Ingo Potrykus and Peter Beyer), had just altered the DNA of rice to boost vitamin A in children to prevent blindness. They called their product Golden Rice, and one of the scientists had ended up on the cover of *Time* magazine.[10] So what was so new?

Wambugu was shaken. As the sixth of ten children in the town of Nyeri, Kenya, she had grown up hungry most of the time. She was the only member of her family to go to university, and as a crop scientist, her mission was to help control the viruses and other plagues that attack crops and remove food from the plate. "When you come from Africa, nutrition, food, is a challenge you experience every day," she said. "In Africa, people are sick because they are hungry, or because their bodies are so weak from hunger and malnutrition. So medicine doesn't work." Yet her colleagues from rich countries didn't get it. "What I thought was obvious was not obvious to everybody. I felt out of place," she later said. "Maybe this is not the forum for me," she told Varmus. But Varmus urged her not to give up.

In any case, giving up was not Wambugu's style. During a coffee break, she photocopied colour maps of Africa that showed the extent of malnutrition among children under age five. She set a map in front of each chair. "For most of these children," she explained to the scientists after the break, "if you give them medication, they won't get well." Only by intervening with the foods that children eat, she said, can we prevent diseases like diarrhea and pneumonia.

Wambugu had nailed the first argument: that there is a tangible link between food and health. Next, she tackled the complaint that this challenge had already been overcome. Golden Rice is a fabulous invention, she told the group, but it fortifies only one crop with one nutrient. What's more, rice is a staple food in Asia, not in Africa. To genetically modify several crops with several nutrients is a challenge of an entirely different order.

Wambugu was making headway. But there were more questions, this time from Elias Zerhouni. How will people in Africa feel about this GM food? Will they protest against it, as Europeans have done? Protests in Europe had already stalled the adoption of Golden Rice. How would people in Africa react to the introduction of genetically modified crops? And what about the concern that, in Africa, there was a lack of appropriate regulation to ensure the safety of GM crops?

"Policy does not operate in a vacuum," responded Wambugu. "There has to be a product." In other words, if policy-makers are looking at a product that makes people healthier, they might not be so quick to dismiss GM food for political reasons. Heads nodded.

There was a final, predictable question from the chorus of scientists: "So what's the critical barrier?" By this point, we were getting worried. Wambugu had raised a crucial issue, but if she didn't grasp the language and the logic of the critical barrier, GM foods would fail as a grand challenge, and research into this important area would not get the Gates funding it deserved. Wambugu wasn't sure how to answer, so Varmus asked Peter to help her draft the language of the grand challenges and the critical barrier. We went to Wambugu's room to plot it out, Peter sitting at a small table in the corner of the room, Florence sitting on her bed. What was the critical barrier? Eventually, we zeroed in on the answer: "We don't know how to put a complex of traits like zinc, iron and vitamin A into crops," Wambugu said. No one had ever stacked three or four traits into a single crop that people in Africa eat every day. No one had put several traits into staple crops eaten by the poor.

We wrote this up in the language of the grand challenges: "We currently lack a strategy for the effective, efficient, and socially and culturally acceptable alteration of a major dietary staple, to achieve the delivery of multiple micronutrients, such as minerals and vitamins, to poor populations in a single food. Moreover, genetic changes could ensure higher protein content and the presence of essential amino acids."

When Wambugu read out our statement to the full group of scientists, Varmus was the first to speak: "Now that sounds more like it." Developing nutritional foods through genetic modification would be Grand Challenge No. 9. Mission accomplished.

Another of the grand challenges we chose in those three days might have surprised observers—but it was no surprise to anyone who knows Julio Frenk, Mexico's health minister at the time. Frenk, a suave and persuasive adviser to both Bill Gates and Mexican billionaire Carlos Slim, is now dean of the Harvard School of Public Health. Yet the idea he was selling wasn't that compelling—at least, not at first. It was about measuring health status accurately and economically in developing countries. It turns out, as Frenk explained, that health statistics, even figures on births and deaths, are just estimates in these countries. Without autopsies, and without vital statistics such as routine birth and death certificates, health workers often can't say for sure what killed a person or how many people are affected by a given disease. This makes it hard, if not impossible, to prioritize scarce health resources in developing countries, and increases the likelihood that politicians make choices based on political whim rather than evidence.

Modern life sciences, Frenk noted, have produced new ways to measure human biology through mathematical tools that analyze complex data. Having access to these tools could help developing countries identify the causes of illness and death. The challenge, as Frenk saw it, was to develop accurate, comparable and convenient technologies and analytic methods that incorporated clinical, biological and behavioural markers for the quantitative assessment of a population's health and the effects of disease. This would help governments in poor countries to set priorities and evaluate interventions.

Frenk's idea sailed through.

Testing individuals on the ground turned into another grand challenge. As we saw in the case of Abdallah's sister Alwiya, the diagnosis of malaria and many other infectious diseases depends on a blood test, which needs to be done in a clinic. It's often hard for people in developing countries to get to a clinic, and the results take time, which delays treatment. But imagine if there were a hand-held device that could diagnose the cause of a child's fever as easily as a home pregnancy test. The power of modern life sciences makes this entirely possible. Such a device could even diagnose multiple conditions, like the Tricorder device Dr. McCoy used on *Star Trek*.

By the end of the three-day session, we had chosen other challenges as well. One would tackle chronic infections that hide in the lymph nodes, such as HIV, hepatitis B and tuberculosis. The big question—the critical barrier—was: how do you lure such infectious agents into the open in order to attack them? Another grand challenge was the thorny medical issue of drug resistance, which arises when disease-causing microbes mutate to evade treatment. People in North America already know an example—penicillin, which has become less effective because it is overprescribed by doctors. In the developing world, resistance is a far more deadly problem. Chloroquine, a common remedy for malaria, is useless in many parts of the world because the parasite has slightly altered itself to escape the drug's intended effects.

On the third and final day, we all posed for a group photo in front of Airlie House. We had chosen fourteen grand challenges in all, and we felt we had made history. Now there was a real prospect that scientists would at last take aim at the terrible infectious diseases that killed so many people in the developing world. But, as Elias Zerhouni would later tell Peter, the grand challenges could have an even more significant ripple effect. They would direct the attention of the world's greatest scientists towards the health problems of the poor. They would, in other words, create an unstoppable momentum for change.

On October 17, 2003, we published[11] the fourteen grand challenges in *Science*. They had been grouped into seven goals: improve vaccines, create new vaccines, control insect vectors, improve nutrition, limit drug resistance, cure latent infection and measure health status. Now the competition began. The Gates Foundation asked for proposals for research projects, with individual grants up to $20 million over five years. By the submission deadline in January 2004, more than fifteen hundred letters of intent from researchers in seventy-five countries had arrived. Just over four hundred of these researchers were invited to make full proposals. By November, the scientific advisory board returned to Airlie Center to approve the projects. They had been selected by Varmus, Klausner and Zerhouni, along with peer review committees for each grand challenge goal.

The 250 scientists leading the forty-four winning projects are an extraordinary group of hotshots, including two Nobel Prize winners: David Baltimore, then president of the California Institute of Technology, is investigating whether immune cells can be genetically engineered to fight off infectious diseases like HIV, while American neuroscientist Richard Axel is studying insects' sense of smell with a view to developing a safe, effective and low-cost insect repellent for developing countries. This was the beginning of a great scientific adventure, to be sure, but would it actually improve global health during our lifetime? No one knew.

Before the grants were announced, the researchers and the Gates Foundation had worked out the milestones and the intellectual property issues to ensure that the developing world could afford the successful projects funded by the Grand Challenges Initiative. Although researchers and their universities, and even participating companies, would hold the patents, the foundation would negotiate deals ensuring that the developing world could get affordable access to each product.

By the fall of 2005, it was time to begin the work. The 250 scientists from the forty-four research teams flew to Seattle for their first meeting with Bill and Melinda Gates. Sitting on stools in front of this eminent scientific audience, the Gates couple described why they cared about global health. Their journey had begun in the mid-1990s, when they read a newspaper article about a disease they had never heard of—rotavirus—that was killing half a million children each year, nearly always in poor countries. "That's got to be a typo," they recalled thinking. "If a single disease were killing that many kids, we would have heard about it because it would have been front-page news. But it wasn't a typo." The brutal conclusion, they said, was that some lives were worth saving, and others were not. "This can't be true. But if it is true, it deserves to be the priority of our giving." Soon after, the Bill & Melinda Gates Foundation directed its efforts to global health, with the motto, "All lives have equal value." The foundation had focused first on delivering vaccines; now it was widening its focus to discover new solutions through the Grand Challenges Initiative.

The conversation was honest, informal and inspiring; both of the Gateses were incredibly impressive. By the end of the meeting it was decided that the grand challenges scientists would meet as a group annually.

In the intervening months and years, it was difficult at first to tell whether the scientists were getting anywhere. Most of them spent the first little while setting themselves up, but before long, the senior members of the Gates Foundation turned up the pressure to produce real results. The pressure came, most visibly, from Dr. Tachi Yamada, the physician and scientist who soon replaced Klausner as head of the Gates Foundation's global health program. Yamada had grown up in war-ravaged Japan, and as a young doctor he had watched a child die in his arms, an experience he would never forget. He eventually climbed the pharmaceutical ladder to become head of research and development at GlaxoSmithKline (GSK) and a member of the multinational's board of directors. Yet he never forgot his commitment to health in developing countries.

By the time the scientists met in Bangkok in 2008, where they were hosted by the Thai princess Maha Chakri Sirindhorn, Yamada was forcefully urging them to hurry up. This highly focused man was clearly changing the grand challenges strategy from straight discovery to "show me the goods." At GSK, he told the assembled scientists, every day of delay in getting a drug to market cost the company millions of dollars. Now, the stakes were much higher: every day of delay was costing lives. The scientists fell quiet. As if to reinforce the point, the room went dark as a series of photographs lit up the screen. These pictures, by veteran war photographer James Nachtwey, showed people dying of tuberculosis and other diseases in the developing world. One black-and-white photo showed a priest in his robes hugging an emaciated skeleton of a man dying of AIDS. When the lights went up, many of the scientists and Gates Foundation staff were weeping.

In the handful of years that had passed since the scientists began their work to solve the grand challenges, some of those who'd received multi-million-dollar grants were already reporting thrilling progress. Stefan Kappe, for instance, is a German-born molecular biologist who is working on an innovative vaccine for malaria. It will not be the first; GSK has developed a vaccine that is now in clinical trials and will likely become the world's first malaria vaccine. Unfortunately, it looks like the GSK vaccine may prevent only a third of cases. But Kappe's novel approach may help close this gap.

Kappe works out of the Seattle Biomedical Research Institute, one of the energy- and water-efficient buildings in that city's fast-growing biomedical hub. His approach to fighting malaria is based on his lifelong interest in parasites—how they have shaped humankind, and how we have shaped them. He's particularly interested in the *Plasmodium falciparum*, which causes the potentially deadly form of the disease.

What makes Kappe's approach special is the strategy he deploys to attack this one-celled killer. When the mosquito bites a person, it delivers a tiny number of parasites—just fifty of them. They swim to the liver and hunker down for a week. The patient feels nothing. Then, after seven days, billions of parasites flood out of the liver into the bloodstream and sicken the individual. "This is the Achilles heel," Kappe explains. The liver, he says, "is the bottleneck where you want to hit them. If you hit them there, people won't get sick."

But how? Kappe is using genetic engineering to disable the parasite just when it counts. Using microarrays, he can tell which of the parasite's genes turn on when the parasite multiplies in the liver. Then he removes those critical dozen or so genes that the parasite needs to grow and multiply. The genetically engineered parasite can still enter the human body but can't cause malaria. When it enters the liver, "it checks in but never checks out. What you present to the immune system is a learning tool." This is an innovative idea that looks like it works. So far, in animal testing, the vaccine appears to be 95 percent effective, and Kappe is now testing the new vaccine on twenty volunteers in the United States. Of course, before his vaccine is used in clinics, Kappe will have to jump through the standard regulatory hoops to ensure it's safe and effective. That takes time, and Kappe doesn't like to wait: "Every day this gets delayed by regulatory and logistical issues drives us crazy."

One way to speed up the path from lab to village is to find a better way to test live vaccines in mice. And one of the leaders in this part of the story is the only scientist outside the developed world to be chosen as a principal investigator in the Gates Grand Challenges. Dr. Hongkui Deng is a professor in the School of Life Sciences in Peking University. When we visited Deng in Beijing, he showed us his ultramodern lab, and then took us, as honoured guests, to a restaurant where the featured act was lavish Chinese opera. "We want to help make an HIV vaccine," Deng explained. "The major problem is that there's no good animal model to use." Mice are the standard model in labs around the world, but the problem with the mouse is that it cannot be infected by the human HIV virus. "We need to replace the mouse's immune system with a human immune system," Deng said. "That would be a much better model."

It would. But how does such a replacement happen? One way is to collect human stem cells from embryos discarded in the in vitro fertilization process and coax them into forming immune-system cells, which can then be injected into the mouse. The supply of human stem cells from discarded embryos is scarce, though, so Deng is using a new technique to take cells from hair, spleen and blood and convert them into stem cells. It's a technical breakthrough, but one that is far from easy to accomplish, let alone repeat. Deng was cheerful about overcoming that obstacle. "Actually, my lab is famous for increasing the technical efficiency up to a hundredfold," he said. Once he creates the stem cells, he induces them, step by step, to become players in the immune system. "It's like sending the kids to school from primary to college," he laughed. Then he injects the new cells into the liver of a newborn mouse. The cells move right into the bone marrow, where they complete their education and become part of the immune system. Only now they're human immune cells—in a mouse. The process is coming along, said Deng. As of the fall of 2009, he had managed to make mice with blood that was 6 percent human, but he was not yet satisfied. "We want 80 percent of the mouse's blood cells to be human."

Although all this work was impressive, we asked Deng how it would help create an HIV vaccine. Creating an effective HIV vaccine is extraordinarily challenging because the virus is constantly mutating. In other words, even if you teach the immune system to recognize one HIV virus, it will not recognize another one that has mutated and looks a little different. The key is to find a virus that causes an immune response not just against that particular virus but against all mutations. Deng explained that mice with human immune systems can help because you can try out many HIV vaccine candidates on many mice to see which one causes the most robust immune response.

Deng is justifiably proud of his participation in the international Grand Challenges, and he is a source of real inspiration. Here is a scientist from a developing country who is making international scientific news on the cutting edge of stem cell research. And he is using science to attack the health problems of the poor, problems that affect his own people as well as people elsewhere in the world.

Deng and Kappe are just two of the exciting researchers who are working on the grand challenge to create new vaccines against the great infectious killers, but their innovations won't do any good unless people can be vaccinated easily and quickly. To spare families repeated treks to the clinic, Dr. Lorne Babiuk, a former director of the internationally renowned Vaccine and Infectious Disease Organization based in Saskatchewan, Canada, is aiming to create a vaccine that can be administered in a single dose to babies just after birth. He spoke to us in the fall of 2009, just as the H1N1 flu virus was making headlines, partly because there wasn't enough vaccine to go around. Two children had died of the virus in Canada, causing panic in many parents. Babiuk was clearly thinking about the far greater number of children who die in developing countries because of the difficulty of travelling to clinics multiple times for shots. It would be so much easier if vaccines could be delivered in one dose shortly after birth, and yet this is a mammoth challenge, he told us. "Children after birth respond differently to vaccines," Babiuk explained. "The mother often transmits antibodies through the placenta or through breast milk that provide early protection, and some of these interfere with vaccines."

Babiuk believes the clue to overcoming the mother's milk barrier lies in the communication system that the body's immune system uses to defend itself. "The immune system is like an army, with ships and planes and subs," Babiuk said. The defensive effort relies on a sophisticated communication system in the body's white blood cells, which passes messages from the front lines to the generals and back again, and alerts the body that an intruder has entered. If the communication system is weak, or is interrupted, the immune system won't be able to mount a powerful defence. If, on the other hand, you enhance that communication system with the help of an adjuvant or booster, the body's defensive system will repel intruders.

Babiuk has boosted the immune system's communication network in baby pigs to ward off whooping cough with just one dose, instead of the customary five. He did this with help from an adjuvant made out of several that are already being used in humans. Now he's nearly ready to test the one-shot vaccine for whooping cough on human infants. "We'll be ready for a trial in one year to eighteen months," he said. If successful, this one-dose routine could be used for all childhood vaccines, and it could even be delivered via a nasal spray.

While these and other successes were discussed and celebrated at the Gates Foundation banquet in Bangkok, another challenge was unveiled, one that expands the program in an exciting way. The grand challenges are big bets, an average $10 million each. Yet they are too big for young innovators, especially from the developing world. So we were delighted to see the Gates Foundation launch a new wave of grants. The Grand Challenges Explorations program expands the pipeline of ideas by offering grants of $100,000, with an extra $1 million to take good ideas to the next stage. This opens the door to exciting young researchers, especially from the developing world, and already they're coming up with some creative ideas, such as using mosquitoes as "flying syringes" to inoculate people, and conducting a study to look at why the eye is so resistant to infection.

The two Grand Challenges programs work together beautifully. While the Explorations grants are mining great ideas bubbling up from the ground, the multi-million-dollar grand challenges projects harness the big wheels of science. By now, the grand challenges teams have made significant progress. Indeed, the scientists who are closest to turning their ideas into reality are the two groups that are altering the language of life to fortify crops that feed children and to disarm mosquitoes that spread disease.

But even if these new scientific technologies prove effective in the lab, we have taken only the first step towards real change in the village. On the treacherous path from lab to village, science can take us only partway. As we will see in the following chapters, scientists have to win the trust of communities to test their innovations and eventually use them. Ideas need to be commercialized, made into actual products and distributed to people in the village. To do this successfully, scientists need to work with business people, something many public health groups have never done before and are not so comfortable doing. And finally, scientists and advocates need to encourage businesses in developing countries to make vaccines, drugs and other health products, and even discover them. This, as we will see, is a major challenge.

CHAPTER THREE

At the beginning of our journey to discover the path from lab to village, we focused almost exclusively on infectious diseases because this seemed the most direct way to save the greatest number of lives. Our driving goal was to break the barriers that block the discovery of life-saving drugs, vaccines and other health technologies. Yet as time passed and we gained some distance from the Gates Grand Challenges, we realized that the very challenges we had helped to identify ignored an important category of disease: those you cannot catch—chronic non-communicable diseases such as cardiovascular disease, certain cancers, diabetes and chronic respiratory disease.

This realization was important because, as we increasingly came to see, these non-communicable killers are devastating the poor world. They kill more people than infectious diseases do in all but the very poorest countries. They're responsible for 60 percent of all deaths worldwide—twice as many deaths as the combined total of HIV/AIDS, TB, malaria, maternal and perinatal conditions and nutritional deficiencies.[1] And they attack old and young alike: nearly half of the people who die of non-communicable illnesses are still in the prime of life, under age seventy. And the saddest part of this picture is that 80 percent of premature heart disease, stroke and diabetes, as well as 40 percent of cancers, can be prevented.[2] They can largely be prevented by eating healthful food, exercising regularly, avoiding tobacco and avoiding the harmful use of alcohol. Yet as the developing world urbanizes, it's actually encouraging non-communicable diseases by fostering conditions that promote obesity—triggering still more disease.

It's no surprise, then, to find that heart disease, stroke, cancers, Type 2 diabetes and chronic respiratory diseases are growing at epidemic rates in many parts of the developing world. What's worse, developing countries are in no position to handle diseases like these. They're too busy coping with the infectious disasters to think about low-cost blood pressure pills, rehabilitation for stroke victims or chemotherapy or even painkillers for people dying from cancer. As a result, in most places in low-income countries, people do not have access to the simplest medicines to prevent or treat diseases like diabetes or conditions like high blood pressure. They have no way of knowing whether they have cancer or other diseases until it's too late. They have little or no access to life-saving procedures or medicines, like heart bypass surgery, catheterization or blood pressure medication that people in rich countries take for granted. Consequently, tens of millions of people die terrible and needless deaths every single year.

And yet these chronic non-communicable diseases have been consistently overlooked by donors and by the governments of rich and poor countries alike. The Gates Foundation, for instance, deliberately excluded them from its Grand Challenges Initiative. For one thing, the causes of these diseases, a mix of genes and environmental triggers, seemed to be too diffuse to tackle. How do you induce a whole country or even a region to stop smoking, exercise, cut back on alcohol and stop feeding sugary drinks to the kids? This is an exceptionally complex task in any country—and even harder in poor ones.

The result is that while we are beginning to make scientific progress in the battle against some of the worst infectious killers in developing countries, people in the villages and cities of those same countries are putting themselves at risk of painful and premature death from non-communicable diseases—either because they don't know they're taking the risk or because they can't afford to live any other way.

We saw a prime example of this contradiction in the fall of 2009 in the village of Usa River, not far from the mountainous city of Arusha, in Abdallah's country of birth, Tanzania. We had just visited a remarkable factory in Arusha that makes bed nets impregnated with insecticide to ward off malaria, a success story we will describe later in this book. As the only native of Tanzania in our party, Abdallah guided our fellow scientists, all high-powered Americans and Europeans, on a bus ride to a village to see how the malaria nets were being used. What we witnessed was an eye-opener.

The bus stopped on a sandy path, and we walked towards the village between small plots of rice and pools of stagnant water. When we arrived at a cluster of mud-brick houses, we found a tiny shop selling lollipops, salt, sugar, some generic medicines and a local brew called Konyagi. Here we met Anna, a tiny woman of indeterminate age in a green-and-yellow batik kanga, the traditional wraparound skirt. Anna showed us around her modest two-bedroom house, where she lives with a daughter, who runs the shop, and four grand-children who were staying with her while their parents worked in another town.

Anna led us towards her small bed, where she sleeps with her four grandchildren. It was covered by a big blue bed net, made by the factory we had just visited. "We don't get malaria now," she said proudly. This one change has greatly improved not only the family's health but their finances. They can work more days in the fields, so they earn more.

"Where do you cook?" Abdallah asked in Swahili.

Anna took us to her cooking hut, a few feet from the main house. As we stepped inside the darkened shack, thick smoke rose from semi-dry twigs burning underneath the cooking pot. Anna was cooking beans and vegetables in a flimsy pot balanced on the fire on three large stones. The smoke stung our eyes and made our throats itch. It was so smoky that we couldn't stay long, so soon we said goodbye and stepped back into the cool outside air.

On the way back to Arusha, we discussed how Anna was doing a fine job of protecting her family from the menace of mosquitoes that spread malaria. But did she realize that she was exposing herself and anyone else in the cooking hut to the long-term health effects of inhaling smoke and soot? Inhaling smoke from indoor cooking leads to chronic lung disease in adults, mainly women, and acute respiratory disease in children. Chronic lung disease causes approximately 1.5 million premature deaths a year, usually among adults.[3]

That indoor fire in the cooking hut illustrates the far bigger problem: in most of the developing world, except for some of the lowest-income countries, people such as Anna are now more likely to die of a non-communicable disease than from an infectious one.[4]

This new picture is only now beginning to emerge, in bits and pieces, from different places in the developing world. What we're learning is that the spread of chronic non-communicable disease is wider than we imagined, more difficult to understand than we know and considerably harder to address than the relatively straight-forward menace of diseases spread by bugs.

India and China, for instance, are now considered the world capitals for Type 2 diabetes,[5] a disease that prevents the body from responding normally to insulin. In India, an astonishing number of people—about 58.7 million[6]—have diabetes. About 85 percent of these have Type 2, a chronic condition that was rare a generation ago. These diabetics face complications such as kidney failure, heart disease, stroke and blindness. Diabetes is now more prevalent in Chennai than it is in New York City. Even worse, researchers are predicting that the number of afflicted could double in the next fifteen years.[7]

This is a looming disaster, because the consequences of getting diabetes in India are far worse than they are in the West. It is esti-mated that more than 100,000 people in India develop end-stage kidney failure each year.[8] For the vast majority of them, it's a death sentence, because dialysis and kidney transplants are unavailable or too expensive, even for a middle-income worker. Doctors in India tell the same sad story: A middle-aged man visits the clinic and is told his kidneys are not working properly because of a disease he didn't even know he had. Then he is faced with an awful choice: use up the family's savings to get dialysis and save his life for a short period, or refuse treatment and die so that the kids can afford to stay in school.

Type 2 diabetes is often called "the sugar disease" partly because sugar is found in the urine, but it might just as well be because the risk of contracting it is heightened by a modern urban diet loaded with condensed calories, which leads to obesity. Of course, the story is not that simple. Studies of Type 2 diabetes show a genetic component to the disease, and so far approximately twenty genes[9] seem to be involved. Yet, as Abdallah discovered on a 2009 trip to India, the problem in developing countries may be even more complex than the combination of diet and genetics may suggest, and may begin before birth.

Dr. Chittaranjan Yajnik is director of the diabetes unit at King Edward Memorial Hospital in Pune, India's eighth-largest city. As Abdallah found, this Indian scientist is shaking up the traditional thinking about the origin of diabetes. He started questioning assumptions about the disease as a student, when he noticed that many diabetics he saw in India did not fit the picture he studied in his medical textbooks. A significant proportion of Indian Type 2 diabetics were short, small and thin—not at all like the fat people who are usually thought to be at risk for the disease. He began researching Indian diabetics and discovered that Indians carried more fat per kilogram of body weight than Europeans did. Although they were not overweight, Indians were "thin but fat." The extra fat was triggering diabetes. "It took some time to sell the idea," said Yajnik, a slender, balding, five-foot-six-inch scientist. His findings contradicted the beliefs firmly held by most scientists, but Yajnik persevered.

Then he made another important finding: the problem begins before birth. When he compared Indian babies with babies born in a prosperous part of the United Kingdom, he noticed that Indian babies were fatter (meaning they had more fat per kilo) than English babies of the same weight. "These are small, thin and fat Indian babies," he said. "People didn't believe it; they thought I hadn't measured right." But the finding was confirmed by sophisticated MRI measurements.

Then Yajnik linked the thin-but-fat condition in Indian babies to their mothers' nutrition. Maternal nutrition programs the body composition and metabolism of the baby. Maternal under-nutrition, he found, makes babies fatter. "Maternal under-nutrition also programs the fetal brain," said Yajnik. "They are more hungry after they're born. They eat more. They tend to deposit more fat and later in life become obese. These children born to underweight mothers are doomed for life."

Yajnik's work is a classic example of the scientists' old line, "Genes load the gun, but the environment pulls the trigger." But his study raises as many questions as it answers. If we want to tackle diabetes on a national or international scale, what kind of food should pregnant women and young children eat? What micronutrients can prevent diabetes? Yajnik can't answer those questions. To reduce the risk of getting diabetes and heart disease later in life, he says, the children in developing countries need better nutrition from the time they are growing in the womb. He thinks a balanced supply of vitamin B12, folate and high-quality protein will be important in India. "But what kind, we don't know." This uncertainty certainly complicates the job of managing diabetes on a large scale.

Meanwhile, diabetes has spread to African countries such as Cameroon, in the western part of the continent. "It's a strange paradox," said Jean Claude Mbanya, one of Africa's foremost authorities on the disease. He is a professor of endocrinology at the University of Yaounde in Cameroon and the president of the International Diabetes Federation. "You'd think people are so poor that everyone will be thin because of the nutritional deficit." Yet Cameroon, like Uganda and several other upwardly mobile African countries, is seeing the downside of urbanization. As they move to the city, people walk less and eat more sugary food as a cheap source of energy, and then many of them get dangerously fat. So now, even as tens of thousands of Cameroonian children in rural areas are undernourished, a substantial percentage of people in the cities—28 percent of women and nearly as many men—are overweight or obese.[10] This change is already driving up the rate of Type 2 diabetes in Cameroon. In the mid-1990s, less than 2 percent of the urban population was diabetic. By the 2000s, this had increased to just under 7 percent.[11] "We have entered a diabetes epidemic," said Mbanya. His comments struck a chord: we heard the same stories when we visited Tanzania, Uganda, Kenya, South Africa and Ghana.

The growing trend of obesity in developing countries is a boon for the world's biggest killer, cardiovascular diseases. Although people think of cardiovascular disease (mainly heart disease and stroke) as a rich man's killer, 82 percent of the deaths take place in low- and middle-income countries.[12] This is where people are more exposed to the known risk factors—unhealthy nutrition (in this case leading to obesity), smoking and lack of exercise. This is where people have less access to drugs to prevent heart attacks, drugs that can cost as little as one dollar a month. This is where governments have no money to spend on the prevention and management of chronic diseases. As a result, people die younger, or if they survive a stroke or a heart attack, they impose a heavy financial burden on their families, who have to dole out a large proportion of their annual incomes for the care of the patient.

One telling story comes from Jimmy Volmink, head of the Department of Primary Care at the University of Cape Town, South Africa. "When I was studying to become a doctor thirty years ago," Volmink said, "our professors were telling students that heart attacks do not occur in Africans. Returning from the U.K. in 1996 with a PhD, I was excited about using my newfound expertise to study heart attacks. My colleagues thought this was the dumbest idea imaginable. But cardiovascular disease is now the number-one cause of death in my province."[13]

Diabetes and cardiovascular disease are not the only killers on the rise. Cancer already causes more deaths each year worldwide than HIV/AIDS, TB and malaria combined,[14] and researchers are now beginning to get a picture of the deaths cancer causes specifically in the developing world. Cervical and breast cancers are common, and lung cancer, in particular, is a growing problem,[15] especially now that Africans are becoming addicted to cigarettes.[16] And yet throughout Africa, millions of people have no access to cancer screening, early diagnosis, treatment—or even morphine to kill the pain at the end of life.

By 2020, about 60 percent of cancer cases will be in developing countries, which are least able to deal with the problem. By 2030, this will rise to 70 percent.[17] Bringing attention to the problem of cancer in the developing world are dedicated scientists in Africa as well as bright African scientists working in the West. One of the latter is Olufunmilayo (Funmi) Olopade, a professor of medicine and human genetics at the University of Chicago. Olopade and her many African colleagues are a ray of hope, part of a growing number of people who are paying attention to, and coming up with solutions for, the vast and complex problem of dealing with cancer in the developing world.

Few had tried to systematically confront the many problems of chronic non-communicable diseases until 2003, just after the Gates Foundation held its historic meeting to choose the grand challenges. Around that time, Abdallah was invited to join a stellar cast of about fifty researchers and academics from many countries who were meeting at Oxford University; they wanted to do something serious about the epidemic of non-communicable diseases in both developed and developing countries. (Abdallah and his colleagues went on to create a new group, called Oxford Vision 2020, which later became the Oxford Health Alliance.)

When the international group arrived, the host, John Bell, Oxford's distinguished Regius Professor of Medicine, welcomed the group to the ancient and magnificent Magdalen College. Bell, a former student of Abdallah's when Abdallah was a lecturer in surgery at Oxford University, is a slim, athletic Canadian who once rowed for Oxford on a Rhodes Scholarship. John had previously studied the genetics of diabetes at Stanford, and from Oxford he has led a consortium that discovered or confirmed many of the twenty genetic variants that predispose people to getting diabetes. So Bell began the meeting by telling us about the genetic part of the story, and what had been learned about how genes interact with the environment. Family studies revealed beyond a doubt that genes play a big role in triggering these non-communicable diseases.

The environment, Bell explained, can switch genes on and off; even food and some drugs can do so. Yet this interplay of genes and the environment is extraordinarily complex, especially in diseases such as diabetes, where many genes are at work.

Soon after our meeting, a major methodological breakthrough made it much easier and much less expensive for scientists to discover new genes associated with diabetes and other chronic diseases. Genome-wide association scans involve rapidly scanning the complete genomes of large numbers of people—instead of just members of one family—and comparing those who have a disease or condition with those who don't to find common genetic variations associated with a particular disease. But we still don't really know exactly what role genes play and how they interact with one another and the environment. Do they work together to make a bigger impact than one gene might have alone? Studies on identical twins and families suggest that the genetic influence on Type 2 diabetes is much higher than what has so far been discovered by isolating the individual genes, so where is this missing heritability, this "genome dark matter"? And if there are genes that increase risk, are there also genes that might reduce risk?

Biobanks may help to shine a light on the mystery of how genes and the environment work together to create or prevent disease. UK Biobank, for example, was launched in 2006 and completed recruiting half a million people by 2010. It will examine their genomic profiles against their life histories to identify genetic factors that contribute to heart disease, diabetes and other chronic diseases and track how these factors interact with the environment. From this information, researchers might understand more about why some people get heart disease or diabetes while others do not. There are several other major biobanks worldwide, such as the European Prospective Investigation into Cancer and Nutrition, set up to look at the link between cancer, genetics and nutrition. In China, the Kadoorie Study of Chronic Diseases is looking at the role of genes and environment, such as tobacco, infections and diet, on premature death and disability. Launched in 1999, the Mexico City Prospective Study is following 150,000 adults over age forty to look at the impact of their smoking and alcohol habits, as well as their diet, on chronic disease. In Iceland, practically the whole nation was part of a biobank initiative that has been seeking to identify genes causing a number of diseases, already with some successes. Estonia, too, has a nation-wide biobank initiative.

These biobanks may eventually help us to understand complex diseases, but answers from the world of genetics will not come any time soon. Can pills help? we wondered. This was a question we discussed at our meeting in Oxford. And indeed, we discovered that pills may potentially provide a fix for some of these problems. Sir Richard Peto, a hero of Abdallah's from his Oxford days, does brilliant work analyzing epidemiological studies that provide the scientific basis for large-scale interventions on entire populations. Peto (like Salim Yusuf of McMaster University in Canada and a few others) often speaks of how a combination of a few inexpensive drugs— Aspirin, an antihypertensive, with perhaps a cheap statin to bring down cholesterol—packaged in a "polypill" could save millions of lives around the world.[18] Such a pill could be especially helpful as a form of "secondary prevention" (following a heart attack, for example) to reduce the likelihood of suffering another attack.

Still, drugs alone can't solve such a vast and complex problem as non-communicable disease. The most obvious way to reduce such diseases is to attack the key drivers—smoking, alcohol, bad nutrition and lack of exercise. But most of the ways to cut risk are complicated and difficult because they involve changing a political and economic and industrial climate that encourages smoking and obesity and discourages exercise.

No one knows this challenge better than Derek Yach, one of the experts who attended the meeting at Oxford. Yach, a star swimmer from South Africa, is one of the few people in the world to have ever swum in the cold, shark-infested waters between Cape Town and Robben Island, where Nelson Mandela was once incarcerated. This courage served him well when, as executive director in charge of chronic diseases at the World Health Organization, and later as representative of the director-general of the WHO, he fought a fierce battle to stop tobacco companies from spreading smoking, and the disease that accompanies it, to poor people in the developing world.

As he would discover, winning the war on smoking took him far beyond the world inhabited by scientists and public health officials.

Yach was on the verge of winning his political battle on the day he spoke to us at Oxford. The stakes were high. While people in rich countries are turning away from smoking, more people are smoking in developing countries. It's estimated that in the twenty-first century, one billion people will die from smoking-related conditions. And if the tobacco companies continue to addict people in the poor world, more than 80 percent of these deaths will occur there.[19]

Yach began early with a fundamental conviction: To persuade large numbers of people to quit smoking, there was no point starting with the smokers themselves. You had to change industrial behaviour, and to do this you had to change political behaviour. You had to persuade politicians to slap taxes on tobacco to raise prices; you had to ban ads extolling tobacco's virtues; and you had to ban smoking from public places. This strategy has worked admirably in rich countries, and Yach contended it could work in poor countries too.

Since tobacco companies operate on a global scale, Yach knew he had to start at the top, at the WHO, in order to achieve a global and consistent line of attack. This wasn't easy. As a medically dominated institution, the WHO had never tried to use international law to deal with the public health disaster that would surely accompany the rise in smoking in the poor world. Yach also had to work on the United Nations so it wouldn't undercut his anti-smoking campaign. He persuaded the World Bank to stop loaning money to farmers to grow tobacco, and to agree that the primary goal of the UN-wide policy on tobacco was to reduce demand. He also helped the WHO support non-governmental organizations to improve their global reach in the fight against smoking and the companies and individuals who support it.

The real test came when Yach and his anti-smoking allies had to persuade reluctant governments—especially the United States, Germany and Japan—to vote for the WHO's first treaty, the WHO Framework Convention on Tobacco Control. It called for tobacco taxes, ad bans, explicit and bold warnings, and bans on smoking in public places. The treaty was adopted in 2003,[20] a huge victory for Yach personally and for the WHO.

By now, 168 countries have ratified the treaty,[21] but implementing it is hard. A law on the books is not the law of the streets. As of 2008, only 5.4 percent of the world's population was covered by comprehensive smoke-free laws—up from 3.1 percent the year earlier, according to the WHO.[22] The good news is that seventeen countries have implemented comprehensive smoke-free laws. The bad news is that everyone else, especially in the developing world, is exposed to an onslaught of tobacco marketing. Not surprisingly, the number of smokers is still growing, from just over a billion people in 2000 to nearly 1.2 billion smokers now. "But even if we can slow the rate of new smokers," Yach says, "we would have a huge impact."

Governments are reluctant to ban tobacco advertising because they think they'll go broke if consumption declines. Similarly, they worry that raising tobacco taxes will decrease sales and thus lower tax revenues. Yet a 1999 report[23] led by Prabhat Jha and published by the World Bank suggests governments needn't worry. "This is one of the most important analytic works on the economics of tobacco," Yach told us recently. "It shows that increasing levels of tobacco taxes won't rob governments of revenue. Even if you increase the price of cigarettes by 10 percent, you can only expect a decline in consumption of 4 percent." Increasing tobacco prices, in other words, will cause government revenues to rise, even as consumption falls. The fall is steepest among the young and poor. As counterintuitive as it sounds, this study may help persuade governments to ban tobacco ads and implement the other parts of the WHO anti-smoking treaty.

Cutting tobacco smoking is a hugely complex and frustrating battle, but it looks simple compared with the challenge of fighting a condition like obesity. There are so many villains in the obesity wars that you can take your pick. You could blame quick-and-dirty urbanization that creates cities where children have nowhere to play and adults have nowhere to walk. You could point the finger at government policy that allows junk food and soda pop to become cheap sources of energy. Or you could blame poverty, which leads malnourished women to give birth to babies who are programmed to become diabetic.

Fighting non-communicable diseases is a daunting challenge, but after taking inspiration from Derek Yach's progress with smoking, the Oxford Health Alliance was not about to back down. We decided to move forward, starting with a £3-million donation from Lars Rebien Sorensen, the CEO of the Danish insulin manufacturer Novo Nordisk. Abdallah undertook a major initiative to design and carry out an international study to identify the grand challenges in chronic non-communicable diseases. But how to go about this? What method would be best? Who would fund the ongoing study?

Abdallah was understandably nervous about the prospect. For one thing, we weren't starting off with a $200-million fund like the one from the Gates Foundation. There was no money to do a major study to identify the challenges, never mind fund the subsequent research needed to address these challenges. How would we motivate and empower the world's top scientific researchers to explore this field? What's more, governments in developing countries were not interested in the idea. Many of them, especially in sub-Saharan

Africa, already had their hands full with fighting infectious diseases. They could barely muster interest about the so-called neglected tropical diseases like schistosomiasis, lymphatic filariasis, hookworms and kala azar, which were harming and killing millions of people.

Abdallah was also skeptical of the role the private sector would play. Food and beverage multinationals are helping to fuel disease by pushing sugary drinks and junk food to schoolkids. Why would they possibly want to help to fix the problem? Having grown up in socialist Tanzania, he had a tendency to see private industry as a corrupting influence in global health, not a potential partner, particularly in addressing chronic diseases. What's more, there were not enough scientists from the developing world in the room at Oxford where we were launching this enterprise. On the surface our mission seemed futile.

Still, we decided to forge ahead with the study. It's worth doing, we thought, precisely because it is so complicated and so difficult. Our previous experience in the studies we had carried out in Toronto had equipped us, perhaps more than any other group in the world, to take on this challenge. In short, we knew it would require incredible resilience—but at the time even we did not appreciate just how hard it would be to interest the world's major players in the biggest killers on the planet.

Our initial goal was to raise $20 to $30 million to fund whatever scientific projects emerged to answer the grand challenges—challenges we had still not identified. John Bell, in his role as one of the most powerful figures in science in Great Britain and as chair of the Oxford Health Alliance, approached some of the major funding agencies in the United Kingdom to ask for money. No luck. Abdallah talked to people in other funding agencies in the U.K. and Canada. "Where's the beef?" they asked. They weren't going to put up money before we could say exactly what sort of major projects the money would fund. They were not, in other words, like Bill Gates, who had done exactly this with his Grand Challenges Initiative.

By the end of 2005, even Abdallah was ready to give up. And then we saw a ray of hope. At that very moment, George Sarna, a senior official of the Medical Research Council in the U.K., came up with an intriguing idea: Why wait for the money? Why not just go ahead and do the grand challenges exercise, and once it was done, use the results to round up the research cash? We had to do something rather than just talk. If the idea failed, it failed. If it succeeded, we would use the resulting grand challenges to raise money.

And so we began with a small grant from the Oxford Health Alliance, subsidized by funds, personnel and expertise from the McLaughlin-Rotman Centre for Global Health in Toronto. Peter joined the scientific advisory board for the study.

We decided to use the Delphi method to find our challenges, the same method we had used in the previous studies. We had shown how it can build consensus among a group of far-flung experts. We invited two hundred scientists from around the world, including many from the developing world, to be part of the study. We asked them a simple question: "What do you think are the grand challenges in chronic non-communicable diseases?" We defined a grand challenge, as before, as a specific critical barrier that, if removed, would help to solve an important health problem. The experts responded with 1,854 proposals. Some targeted scientific barriers, but most pointed to issues of policy or politics. This time, clearly, the answer did not start in the lab. Over three rounds of questions and answers, we crystallized the ideas into twenty grand challenges that we grouped into six main goals. We also suggested thirty-nine research projects that could address these challenges.[24]

The first and main grand challenge was to raise political awareness and priority for non-communicable disease, especially among policy-makers. This would be crucial. If politicians didn't get excited about the topic, they wouldn't vote for budgets that would make a difference or develop long-term policies with no immediate political payback—such as promoting healthy lifestyle and consumption choices. Of course, this would be a tall order for governments of poor countries. Ask any health minister about attacking non-communicable diseases and you're likely to get the same response: "You must be joking! I've got fifteen dollars per person per year to pay for hospitals, drugs and all the other costs, and you want me to start thinking about curbs and parks? Go to the minister of planning. I have other things to worry about, like kids dying of malaria and the HIV epidemic." So the challenge here was how to persuade politicians and policy-makers to pay attention to the long-term threat without taking attention away from the immediate demands of infectious disease.

Once politicians in developing countries could be made to focus on the issue of non-communicable disease, they would have to consider a broad array of policies that affected health—in particular, economic, legal and environmental policies. This was the second major goal. For instance, how could governments develop trade agreements that discouraged the consumption of alcohol, tobacco and unhealthy foods? How could they develop procedures to evaluate agricultural policies—such as what food to grow—in terms of their effect on health? Should governments, for example, insist on food labels? Were such labels even effective?

The third major goal was to reduce the risk factors that lead to non-communicable disease. For example, we would call on countries to cut tobacco use by implementing the WHO Framework Convention on Tobacco Control. Reducing risk factors could mean many other things as well, such as increasing the availability of healthy foods and promoting lifelong physical activity. This could lead to several important research projects, such as prospective cohort studies to identify the factors that magnify or reduce the risk of getting disease, as well as a systematic multi-country study of the influence of fetal and early-life nutrition on the risk of getting chronic disease. Another research project could investigate the cultural and ethnic variations among risk factors. The results could help governments refine and adapt behavioural interventions.

Was the prevention and treatment of non-communicable disease a government's job exclusively? Perhaps not. Our panel decided we needed to engage business as a partner in promoting health and preventing disease. To achieve this fourth goal, we would study business marketing techniques and learn, for example, how taste, flavour, packaging, labelling and advertising affect consumers' food choices. We needed to build trust with industry and to develop codes of responsible conduct.

To many people who work in public health, engaging business in the promotion of public health is a surprising goal. They see public health as a government job, not one suited to profit-minded business. Yet one of the major proponents of this strategy is Derek Yach, who led the WHO campaign to cut smoking worldwide. In 2007, he surprised the public health world by taking a job at PepsiCo, the maker of sugary pop and salt-and-fat-laden potato chips, products that account for three-quarters of the company's sales. Now Yach, who is vice-president of global health policy at PepsiCo, is helping the company to fulfill the promise of its chairman and CEO, Indra Nooyi, to improve the nutrition of its foods and beverages by cutting their salt, sugar and fat content. In addition, artificial sweeteners are being replaced with natural ones. PepsiCo has taken the lead among others in the industry in reducing the flow of high-calorie drinks and snacks to schoolchildren in the United States and has promised, along with the Coca-Cola Company, to stop advertising high-calorie sugary drinks to children under twelve worldwide—including in the developing world—and to eliminate the direct sales of full-calorie sodas from schools globally by the end of 2011. (PepsiCo will do so in all schools, whereas the Coca-Cola Company will do so only in junior schools.)

These are voluntary deals, and Yach argues that they should stay that way. Governments of poor countries, he says, do not yet have the means to regulate or enforce changes to the beverage and food industry, which has far more powerful players than the tobacco business. Instead, he thinks government should set the rules and leave it to the industry itself to spot the violators. Coca-Cola and PepsiCo will have an incentive to do so, if only to keep the playing field even. The beauty of self-regulation, according to Yach, is not only that industry has to pay for it but that it can be readily adapted to the realities of specific countries. "It's much better than relying on weak government regulatory capacity. Furthermore, all pledges will be subject to independent audits. The school pledge, for example, will be monitored by the World Heart Federation." Even so, we will continue to be skeptical until we see it implemented successfully.

The fifth goal we identified is the most profound one: to mitigate the health impacts of poverty and urbanization. How does poverty increase risk factors? What are the links between the built environment, urbanization and chronic non-communicable disease? To answer these questions, researchers could investigate the biological basis for how poverty raises the risk of disease. We could also study how to work with planners, architects and city politicians to make cities in the poor world healthier places to live.

The final goal was perhaps the most ambitious one: to reorient health systems to take prevention more seriously and to include non-communicable diseases in their mission. To achieve this goal, we suggested basing the resources to be allocated on the burden of disease, rather than on the whim of politicians and donors. In many developing countries, health professionals often work by responding to crises, mostly those involving infectious diseases. We wanted to discover practical ways to get more of them to work on prevention and the management of chronic diseases. And we wanted to know: can health services learn to integrate the management of both infectious and chronic non-communicable diseases?

Clearly, this set of grand challenges would be dramatically different from the grand challenges we had helped identify with the Gates Foundation. While the Gates initiative focused squarely on disease itself (and specifically infectious disease), our new grand challenges aimed to integrate science, technology and policy to create a comprehensive and demanding blueprint for change on a grand scale.

We achieved our first victory in November 2007, when our paper based on the study of grand challenges in chronic non-communicable diseases was published[25] in *Nature*. Our list of grand challenges, conveniently assembled in a one-page table, had a huge and instant impact internationally. Less than two years later, in June of 2009, six of the world's foremost health research funding agencies met in Seattle to announce their collaboration in the battle against non-communicable diseases. The Global Alliance for Chronic Disease, as we called it, would be an important new way to coordinate approaches to funding research and avoid duplication as we try to understand how to manage this terrible epidemic of chronic non-communicable diseases around the world. Considering where we had started, this was a magical outcome.

The health agencies that make up the Global Alliance have agreed to collaborate on worldwide research projects that focus on how to implement ideas on the ground, with a specific focus on low- and middle-income countries. Since the original six members of the alliance collectively manage an estimated 80 percent of all public health research funding around the globe, it was conceivable that they and the new members could one day invest as much in chronic non-communicable diseases as the Bill & Melinda Gates Foundation does in infectious diseases.

Abdallah agreed to chair the alliance, which then moved quickly. At a meeting in New Delhi that November, the alliance announced its first research targets. There are three: lowering high blood pressure (hypertension), reducing smoking of cigarettes and other types of tobacco, and reducing the indoor pollution caused by crude cooking stoves in developing countries—stoves such as Anna's cooking fire in the hut we visited in Tanzania. Together, these three problems contribute to about one in five deaths worldwide each year. And this is only the beginning. We want to try out pilot projects, and if they work effectively on one condition or disease, we want to use the same techniques on others such as diabetes.

At Abdallah's instigation, the alliance also agreed to launch an even bigger study, this one of the grand challenges in global mental health. Mental health is largely ignored around the world, and particularly in developing countries. It is not traditionally listed with other chronic diseases, and yet we decided to include it partly because of its link to other non-communicable diseases and the rising rates of death from mental disorders and distress—for example, one million suicides annually, as well as substance abuse and alcoholism leading to death by illness and injury. Abdallah agreed to lead this study together with Vikram Patel of the London School of Hygiene and Tropical Medicine, and Pamela Collins and Tom Insel of the U.S. National Institute of Mental Health. Peter is on the scientific advisory board. By September 2010 this study was well under way, and we expect to complete it before the end of 2011.

One key issue for our alliance is how to implement scientific innovations in "the village"—or as we have put it elsewhere in these pages, how to transport scientific innovations from the lab to the village. Getting people in the village to try a new drug or adopt a new behaviour has been a massive challenge, and many scientific ideas have failed along the way—or have helped far fewer people than they should have. A classic example is hypertension. In Tanzania, about 30 percent of all men and women in a middle-income urban district of Dar es Salaam and a village in the relatively prosperous rural area of Kilimanjaro were found to be hypertensive. Yet fewer than one in five of these knew they had high blood pressure, only 10 percent of them had received any treatment, and in less than 1 percent of those cases was the high blood pressure controlled.[26] These low levels of detection, treatment and control are likely representative of much of the developing world. So the incidence of stroke and the toll from complications is huge. What's more, these strokes are happening in countries without rehabilitation facilities to deal with the victims. In China, for example, stroke is one of the biggest killers, if not the biggest killer, of the adult population.[27]

One problem, as Roger Glass (one of the co-authors of our *Nature* paper) and his colleagues at the John E. Fogarty International Center at the U.S. National Institutes of Health pointed out, is that implementation research has not, until now, been considered a serious science. While randomized controlled experiments are considered the gold standard for testing the safety and efficacy of pharmaceutical drugs, "health delivery schemes are less likely to be subject to rigorous scientific analysis."[28] In the village, this means that people living in poverty often have limited knowledge of how to prevent disease and poor access to quality health care.

Yet now—with the research funding made possible by the Global Alliance for Chronic Diseases, which is a direct result of the grand challenges initiative for non-communicable disease—the newly energized science of implementation is emerging to focus on this very gap between lab and village. It will seek to answer questions such as: Why do some innovations fail in the real world? How can we deliver innovations from the lab to the village in the most cost-effective way?

The first studies are starting to appear, and we at the alliance are hoping to fund more. Take, for example, the Tanzanian cooking hut we described at the beginning of this chapter. It turns out that in neighbouring Kenya a ceramic stove, the Kenya Ceramic Jiko, was developed by local and international researchers and has been available for many years. It reduces charcoal use and pollution, but even in Kenya it reaches less than 20 percent of the rural population.[29] So how do you get a woman like Anna to use such a stove? We'd have to answer a great many more questions: What are the cultural dimensions of this practice? What about the engineering and design issues we need to solve to make an even better stove? What's the cost going to be, and what kind of business model will provide incentives to make these less-polluting stoves and distribute them? How might they be made available to the greatest number of families, and how do we get those families to actually use them? And if such a program works on a small scale, how can it be rolled out on a national scale?

This last question anticipates one of the biggest challenges. We know how to put a single person on a diet or a smoking-cessation program. But how do you develop such a diet or program for a whole society? That's the vital question that implementation science, a new field in global health research, hopes to answer.

CHAPTER FOUR

So far in our journey to bring the most exciting and potentially beneficial scientific and medical advances to the developing world and developed world alike, we have been looking closely at the process of setting goals and defining scientific research. In this chapter, we are going to take the next step: we'll turn our attention to investigating how the results of research are delivered, and how they might best be delivered in the future. What we discovered—perhaps inevitably—is that on the long road from the lab to the village, scientists are bound to encounter plenty of roadblocks that can stop them in their tracks. The trouble often begins at the very beginning of the journey out of the lab, when scientists do a clinical trial to see whether their innovation, be it a drug, vaccine or other technology, is reasonably safe and effective. The trial must respect the latest ethical standards, which are constantly evolving. If this sounds straightforward, it is not—especially when the innovation to be tested is controversial.

Consider the fate of the oral antiretroviral drug tenofovir. In 1995,[1] laboratory tests on monkeys suggested that it might be able to repel the virus responsible for HIV and prevent this deadly disease. Now here was fantastic news: at last we had an innovation that could potentially save millions of lives in the developing world, where HIV was spreading like wildfire. That year, there were 4.7 million new HIV infections worldwide, an average of 13,000 new infections each day. Of these, 2.5 million occurred in Southeast Asia and 1.9 million in sub-Saharan Africa.[2] The strategy behind the innovation was thoughtful too: give the power to prevent HIV to women, especially those at high risk of infection, so that they wouldn't have to depend on men's use of condoms to protect themselves.

The new drug was developed by the California biopharmaceutical company Gilead Sciences.[3] The plan was to test it on a group of women who ran a high risk of getting HIV—female sex workers in the slums of developing countries. Investigators flew to Phnom Penh, Cambodia, to recruit 960 women for a randomized clinical trial.[4] It would compare the rate of HIV infection in women who took the oral medication with the rate in women given a placebo over a period of twelve months. The trial was funded by some heavy hitters, led by the U.S. National Institute of Allergy and Infectious Diseases.[5] Yet as these sophisticated and well-meaning scientists would soon find out, the demands of running a clinical trial in this community could thwart even the best intentions.

The trouble began in Bangkok, in the summer of 2004, at the 15th International AIDS Conference. Members of the Paris branch of

Act Up, an international protest group concerned with AIDS issues, gathered in front of the Gilead booth to show TV reporters signs saying "Sex workers infected by Gilead" and "Tenofovir makes me vomit."[6] The activists charged, among other things, that the women participating in the trial were not getting long-term health insurance to handle possible side effects of the drug. The protesters eventually closed the Gilead booth.[7] Then, as the controversy heated up, Cambodian prime minister Hun Sen[8] stepped into the fray. "If a trial is needed," he said, "please do it on animals, and don't use it on Cambodians."[9]

When we read the news, we were deeply alarmed. Without yet knowing all the details of what had been done or not done, we felt cancelling the trial would be a real tragedy, especially for the women in Phnom Penh. Here was research that could have given necessary and concrete help to women in the developing world, particularly sex workers, a disempowered and vulnerable community, and yet the trial was shut down with vociferous support from both local and overseas activists. We found it hard not to interpret this as a case of the activism of the rich denying health to the poor. How ironic: some of the same activists who routinely complain about multinational companies imposing control on the developing world were now themselves seemingly imposing their agenda and their values on the developing world—apparently with little regard for the basic health of the people who would be denied the benefits of the latest scientific research.

That was our first reaction. But then we explored more deeply and as objectively as possible the ethical issues that arise from the tenofovir story as a kind of test case. We found the tenofovir case to be even more complicated and nuanced than we had imagined. There are no heroes and villains in this story, partly because the ethical standards governing clinical trials, especially those engaging in research involving communities such as that of the sex workers, were shifting quickly. But one truth emerged clearly from our investigation: if scientists neglect the broad ethical dimensions of their clinical trial, especially when working with vulnerable communities in the developing world and in controversial areas like sex, their research will be shut down and they will never achieve their scientific goals. As we discovered through this example and others that followed, it is crucial to engage the community in the process. The community needs to not only support the research but to demand it, especially when the research is controversial. Without that support, even the most scientifically brilliant technology will never be used in practice. Science, in other words, is necessary, but science alone is not sufficient to take an innovation from lab to village. Respect for ethics, and especially the ethical need to have community engagement when testing or delivering potentially controversial technologies, is crucial.

As we considered the tenofovir case, we realized that the broader ethical issues would soon become serious ones for the Gates Foundation. Here it was, awarding over $400 million to investigators trying to solve the grand challenges that we had helped to identify. But the exercise would not fulfill its potential if successful innovations never made it to the village. We believed we could help Gates-funded researchers overcome these barriers on the road from lab to village, and the Gates Foundation agreed, giving us a $10-million grant to help all forty-four teams of investigators in the Grand Challenges in Global Health Initiative address the ethical issues raised by their research.

We thought carefully about what our role would be. We didn't see ourselves as police dogs; that was a job for regulators. Nor did we see ourselves as lap dogs; ethical window dressing would certainly not help investigators if they ran into a tenofovir-style controversy. Rather, we aimed to collaborate with investigators to make sure their research had a sound ethical footing, especially with respect to any novel or unforeseen ethical issues they might encounter. This was an unusual role for bioethicists, who are more used to criticizing the work of researchers than working with them to solve problems. But if life-saving drugs, vaccines and other technologies were to make it to the villages in the developing world, we knew we'd have to help the researchers conduct their research in the most ethical way.

We knew, of course, that research in the developing world with vulnerable communities, on controversial subjects or using controversial technologies, was an extraordinarily sensitive issue. A decade earlier, the scientific world had been rocked by a series of scandals over the treatment of people who had participated in clinical trials in developing countries. A prime example was exposed in a series[10] called "The Body Hunters" that was published in December 2000 by the *Washington Post*. During a 1996 epidemic of bacterial meningitis that swept through Nigeria and killed over fifteen thousand people, a researcher from Pfizer, the world's biggest research-based drug company, saw an opportunity to test a new antibiotic, Trovan. A Pfizer team flew to Kano, a city in northern Nigeria, to test Trovan on infected children. Health workers in Kano already had a cheap intravenous antibiotic called chloramphenicol to treat meningitis, but the Pfizer team wanted to see whether Trovan, in pill or drink form, worked just as well. Two hundred children in Kano were enrolled in the study. Half were given Trovan; the other half were injected with an established antibiotic. After two weeks, five children died in the Trovan group and six in the control group, an insignificant difference. That suggested the new drug, which could be given orally, was just as effective as the established injectable one.

Upon closer inspection, it turned out, as the *Post* reported, that the trial was riddled with many ethical problems. For one thing, rather than testing Trovan against chloramphenicol, the antibiotic in use in the community, it was tested against a more expensive alternative, ceftriaxone. The families did not sign written consent forms, and the lead Nigerian researcher may have lied about getting the approval of a hospital ethics committee in advance.[11]

This kind of scandal used to happen in the United States back in the 1950s and 1960s. In the infamous Tuskegee study, for instance, antibiotics were withheld from African American men with syphilis. The subsequent outcry led to the National Commission on the Protection of Human Subjects of Biomedical and Behavioral Research, and the imposition of strict rules for clinical research. Now the same problem had arrived in the developing world, where many of the world's clinical trials are now done. This leads many experts to wonder whether clinical trials should be conducted in developing countries at all if the drugs being tested are for rich people in rich countries. Marcia Angell, the former editor in chief of the *New England Journal of Medicine*, is an eloquent advocate for this point of view: "The only clinical research that clearly needs to be conducted in the third world is research on third-world diseases," she wrote in an article[12] in the *New York Review of Books*. "Regulations governing research in poor countries should be every bit as stringent—and enforced just as vigilantly—as in well-to-do countries. There is no justification for the present situation in which the standards are looser precisely where human subjects are most vulnerable."

Angell makes an important point, but of course the world on the ground is not so black and white. As we saw in the last chapter, the concept of "third world diseases" is not always so clear. Noncommunicable diseases like heart disease and cancer, though normally thought of as rich-world diseases, in fact kill more people in the developing world than do infectious diseases. Moreover, the risks of testing a new technology can look very different in the developing world than in the rest of the world. Take, for example, the rotavirus, which causes diarrhea and kills half a million children under five years of age in the developing world every year.[13] An early vaccine manufactured by Wyeth Lederle Vaccines was abandoned in 1999 because clinical trials in the rich world detected a rare but serious side effect.[14] Cancelling the vaccine trial made sense in the rich world when you consider how few deaths rotavirus-induced diarrhea causes there. But in the developing world, when you weigh a handful of deaths from the vaccine against the annual toll of half a million dead children, the decision makes considerably less sense.

Research, with the right conditions can be good for the developing world. It improves clinical care as well as a community's scientific and technical capacity. It can even improve a community's economic well-being. While it's true that ethical standards and processes of research in the developing world are not as rigorous as they are in the rich world, this should not be a reason to deny poor communities the benefits of research. Instead, ethical standards need to be raised, and their proper implementation needs to be handled with greater care. Among other things, the developing world needs to gain the expertise to interpret and apply ethical rules. This is one of the key reasons why we started our training program in bioethics for people from the developing world.

How, exactly, can research be done ethically and in a way that respects communities in the developing world? This was the question we asked ourselves, and to find the answer we probed more deeply into the tenofovir case. What we found was that the tenofovir story was not one of poor people being used as guinea pigs to test a drug for the rich; tenofovir could have helped sex workers in the developing world. Yet the very people whose fellow citizens could have benefited from the research rejected the trial. Could this failure have been avoided?

We turned to one of our colleagues to find out. We first met Jerome Singh when he came to Toronto from South Africa to earn his master's degree in bioethics. He was one of the students from the developing world we were training to oversee the ethics of clinical trials in their own countries. Jerome's education was funded by a grant we had received from the U.S. National Institutes of Health's John E. Fogarty International Center, which runs a program to strengthen research ethics capacity in the developing world. (This program had been set up partly in response to the *Washington Post*'s "Body Hunters" articles.) Singh has a sharp eye for the politics of oppression. He grew up in South Africa, and as a "coloured" person, he attended a racially segregated school that couldn't afford to repair its broken windows. Despite this, Singh earned his doctorate in law at the University of Natal; for his thesis he investigated the South African apartheid government's secret program, "Project Coast," which had sought to develop biological and chemical weapons. The program had bold ambitions, such as to develop bioweapons that would harm only blacks. Thankfully, these objectives were not achieved, but they show how science can be misused.

When Singh studied the tenofovir failure, his findings surprised us. Many of the activists' complaints, it turned out, were unfounded. The scientists conducting the trials in Cambodia *did* obtain the consent of the sex workers. They used an informed-consent document written with help from local participants. Investigators also gave proper counselling to the participants and urged them to encourage their customers to use male condoms to prevent the disease. They didn't encourage the sex workers to have unprotected sex. And contrary to the activists' complaints, the tenofovir investigators promised participants that if they got sick, they would get state-of-the-art antiretroviral therapy, in accordance with WHO criteria for the treatment of AIDS, with the possibility of extending the treatment after the trial ended.

The activists, however, were right in a couple of their complaints. The trial's sponsor did not offer the participants long-term health insurance, although some people would consider this demand unreasonable. (It's not required, even in rich countries.) And although the investigators did reach out to some of the community leaders, they didn't fully involve the broader community in Cambodia. They did not, for example, set up community advisory panels to deal with questions and concerns. In fairness, this wouldn't have occurred to most researchers at the time; it wasn't the standard, and might not have seemed necessary. We now have a much greater understanding of the need for such panels, and standards in this area have therefore changed.[15]

Indeed, one of the key things the tenofovir story shows us is that individual consent in the developing world is not enough. We remembered the words of Cameroonian philosopher Godfrey Tangwa, whom we had interviewed in 2007: "African countries, they are communal societies, and scarcely do people act like individuals," he said. "So it's always good to start at the community level because people feel surer of what they are doing if they know that their community knows about it and approves of it. You need to get people involved at the grassroots, but this means coming close to them," which many researchers, many scientists, are either unable or unwilling to do. Tangwa was obviously not talking about simply getting someone to sign an informed-consent form, which can be done without really understanding what such a form says or means. "It means coming close to them and therefore being able to communicate and interact with them, to know what their own ideas are, what their needs are," he continued. "Some of these diseases are diseases they have suffered from for decades and therefore they have very definite attitudes towards these diseases."

People need to be involved in a clinical trial from the design stage on, Tangwa insisted, and engaging them from the very beginning may be the most important thing that a researcher does. As we were to discover as we delved deeper into the tenofovir trial, the key problem is that researchers don't know how to involve communities in the developing world because the question of how exactly to do so—so vital to the proper and successful conduct of research—has never been carefully examined.

Godfrey Tangwa's words should be read by every researcher doing studies in the developing world. He's talking about true community engagement: building a deep and lasting relationship with people at the grassroots level. These are the people who are suffering from the diseases researchers are studying, and therefore they should be the driving force that propels the research. They are ultimately the ones who care most about the outcome of the trial, and if they become full partners in the research, they can help researchers in many ways, especially if the trial enters controversial territory.

We had interviewed Tangwa as part of a study[16] we conducted in 2007 to learn what forces determine whether innovative technologies like genetically modified foods and anti-HIV drugs make it from lab to village. We interviewed seventy experts in the developing world from academia, industry, government and civil society, and what they told us was enlightening, especially for anyone who wants to make a real difference in the village. The gist of what they said is this: You have to engage the local community or the general public to agree to test the new product and eventually to adopt it. You have to make sure it's culturally acceptable: for example, people in Africa were reluctant to use white bed nets for malaria because that was the colour of death shrouds.[17] Being sensitive to cultural nuance is the only way to win the trust of a community in the poor world where a drug, vaccine, diagnostic or device is to be tested. Engaging the community, in other words, is not only the right way to do things; it's the only way to do things.

Now, in studying the tenofovir case, we wanted to see an example of what successful community engagement looks like. Fortunately, Jerome Singh knew where to take us. After earning his master's in bioethics at the University of Toronto, he established the ethics program at the Centre for the AIDS Programme of Research in South Africa, or CAPRISA, an HIV research centre based in Durban. CAPRISA, he told us, had been helping a rural community in South Africa deal with a disastrous outbreak of HIV. Against all odds, this beleaguered community was doing what sex workers in Cambodia refused to do—try out a gel developed in California that might protect them from the disease.

Vulindlela, a rural community of about four hundred thousand, is 150 kilometres from Singh's home in Durban and a world away from that city. It lies in a lush green valley at the end of a dirt road that seems to melt into the land. As Jerome and Peter drove along the barely visible road, past the traditional thatched huts where subsistence farmers and migrant workers live, he explained the latest chapter of the town's story.

Back in 2001, the community noticed that an unusual number of teenagers were dying of HIV, the disease that was ravaging the country. Vulindlela's two traditional chiefs, Nkosi Sondelani Zondi and Nkosi Nsikayezwe Zondi, were growing increasingly alarmed. No one in authority was trying to control this health disaster. The teenagers' lives could have been saved by powerful antiretroviral, or

ARV, drugs, but these were far too expensive for most people. Meanwhile, the president of South Africa, Thabo Mbeki, was expressing his doubts that HIV caused AIDS, while his health minister, Dr. Manto Tshabalala-Msimang, was prescribing a diet of garlic, lemon juice and beets instead of "toxic" ARVs.[18] This perfect storm would soon lead South Africa to have the world's highest number of cases of HIV.

Yet the chiefs in Vulindlela were determined to fight this infectious killer with the power of science, not the hocus-pocus advocated by South Africa's health minister and supported by the president. They asked CAPRISA to come and help. The centre's co-director, Dr. Salim Abdool Karim, could sense the scope of the disaster as soon as he arrived in Vulindlela one Saturday. The traditional leader who welcomed him apologized that he could not spend the entire day as planned with the researchers—he had several funerals to attend.

Over the next few years, CAPRISA found that the prevalence of HIV was rising at a frightening rate. Annual surveys of the people attending the antenatal clinic showed that the rate of HIV infection jumped from 26 percent in 2001 to 43 percent just two years later. Over half of women in their early twenties had been infected with HIV.[19] The devastation implied by these figures is almost unimaginable to those who haven't experienced it first hand.

The sheer numbers did offer one ray of hope, though: there were enough young women at high risk of getting HIV in Vulindlela to test an experimental antiretroviral gel that might save their lives. This gel had the same active ingredient, tenofovir, as the pill that had been rejected by Cameroon. CAPRISA's trial would evaluate the safety and effectiveness of the gel, in a classic randomized trial comparing what happened to women who used the antiretroviral gel with those who used a placebo gel.[20]

As part of this effort, Singh's job was to ensure that the trial would comply with international ethics guidelines and the law. He soon hit his first major obstacle. This was Zulu country; the entire topic of sex was taboo. Although a significant percentage of the community was HIV-positive, very few people publicly admitted this. In addition, there was no word in Zulu for *research*. It was a foreign concept.

Faced with this obstacle, the researchers spent considerable time listening to and talking with the community. They set up regular meetings, which often took place on a pleasant porch near the community centre. They built a gym and encouraged the kids to play in the yard while the adults talked. "We had to create a lexicon to find appropriate words to discuss the research," said Singh. Then they organized small groups of women to encourage them to talk more freely. Finally, they began to open up. It turned out that some of the teenage girls were trading sex with older men for rides to town or to school, or for a cellphone, a dress or other favours. In a town with high unemployment and no prospects, they saw no other choice.

By the end of the discussions, the women were asking astute questions about the gel. If this is used for sex, one grandmother asked, how will it be effective if people don't plan in advance? What's the point? Or this, from another elderly woman: How do you know the gel works if you're asking women to tell their partners to wear condoms? This was a good question, showing real understanding of the concept of research. The women asked about access to ARVs if they became sick. (Women who got infected and were eligible for treatment would be offered CAPRISA's treatment program.) In the end, 889 women in Vulindlela and Durban enrolled in the trial.[21] But the story didn't end there.

The stakes for this clinical trial were high. The research results for microbicides—gels, creams or foams applied to the vagina to prevent HIV transmission—had so far been dismal.[22] This trial was different because the gel's main component was tenofovir, an antiretroviral already used in the treatment of HIV. This, then, would be the world's first human study of a gel based on tenofovir. If it worked, it could potentially lead to a treatment that could save millions of lives. The researchers had to get it right. Based on their knowledge of the realities in the community, they altered the standard clinical trial protocol in one significant way. They saw no reason why the women should apply the vaginal gel every day if their husbands were away working or if they didn't expect to have sex.[23] Why not apply it only before sex? Salim Abdool Karim, co-director of CAPRISA, thought this was a good idea. Yet it provoked a controversy among scientists around the world.[24] These scientists said the results would be worthless if Abdool Karim altered the standard protocol of regular application. Abdool Karim stood his ground. The message to the world's scientists was plain: Vulindlela will do it their way. The gel study went ahead.

As the trial began, researchers were soon listening in on a troubled and sometimes violent world. This was unusual in a clinical trial, but these researchers were extraordinarily committed to listening to the communities they served. One day at a meeting, they learned that one of the participants had been beaten to death by a boyfriend. For CAPRISA researcher Koleka Mlisana, that death raised some hard questions: "What is our obligation? Can we just turn our backs?" CAPRISA researchers often show women videos about domestic abuse, but these tools have had little impact. According to Mlisana, "They always say no, no, no, we always fight when we're drunk and it's not a big issue, you know, life just carries on the next day." It was upsetting, and Mlisana was uncertain what she, as a researcher, was supposed to do to help these women protect themselves. "I told

Jerome we can't just accept this, something has got to be done," she said to us. "It's our responsibility to get the women counselled." But for the women, she found, that wasn't the priority. "As far as they are concerned, this man is not abusing me. He's given me a place to live, he's given me food and that's all that matters to me."

Mlisana raises an important question. Activists have been criticizing these kinds of trials because they suspect that researchers are encouraging participants to have risky sex so that they can clock the research results. No one in Vulindlela was encouraging the women in the study to have unprotected sex. But did the researchers have a duty to stop the women's risky behaviour?

The same question had come up before, in a trial conducted in the slums on the edge of Nairobi. Frank Plummer, a medical microbiologist at the University of Manitoba, went to the slum in the early 1980s to study female sex workers there. The researchers' initial goal was to study sexually transmitted infections with names few North Americans have heard of, names like chancroid and lymphogranuloma venereum. Then the HIV epidemic erupted, and researchers discovered that two-thirds of Nairobi sex workers were HIV-positive—a controversial finding in Kenya, which insisted at the time that it didn't have a problem.

But then Plummer discovered something that startled the scientific world: some of the women in these slums seemed to have a natural immunity to HIV. The longer they were sex workers, the less likely they were to be infected with the virus. What's more, natural immunity seemed to run in families; sex workers related to sex workers with natural immunity were ten times less likely to be infected with HIV than other sex workers.

"We think this is a model of natural immunity to HIV," Plummer told us when we called him in Manitoba, where he is now scientific director of the Public Health Agency of Canada's National Microbiological Laboratory. "It will help us to develop the HIV vaccine." The big question, though, is why some people are immune to HIV while others are not. There are several possibilities, Plummer explained. Immune women might have received a low dose of HIV and become immunized that way. They might possess a genetic profile that protects them; several genes are targets of investigation right now. Whatever the reasons, the immune systems of these people are strong enough to fend off HIV infection. The explanation likely has to do with a combination of several complex factors, so now Plummer is using advanced systems biology to understand the complicated interactions that may create natural immunity. (His research is funded by a Grand Challenges grant from the Gates Foundation and the Canadian Institutes of Health Research.)

Plummer can do this research only because a group of sex workers in Nairobi have agreed to give him blood samples and have participated in the research over the past twenty-five years. This long partnership began when Plummer met with the district chief in the Nairobi slum where the women worked. Plummer made it clear that these women would get free health care, counselling and condoms. Five hundred women registered on the first day.

Then the community organized, and elected representatives to bring forward problems and discuss issues. "It was really empowering for the community," said Plummer. "Prior to us being there, there was zero condom use. Men didn't like them. Then the community decided to insist on condoms." Now, women say they're protected by condoms during 80 percent of sex acts. (Condom use didn't affect the research because unfortunately the rate of HIV went up so fast that the exposure to the virus for the sex workers remained constant.)

This is a successful partnership between the women and the scientists, but it has still raised some serious questions for journalist Stephanie Nolen, who wrote about the hard life of one of the sex workers in her book[25] *28: Stories of AIDS in Africa*. Agnes Munyiva, a sex worker profiled in Nolen's book, has a rough and unrelenting life. Every day, she gives sex to a dozen men, for a dollar or two each time, on a hard, lumpy mattress in a mud-walled room, where the air "has a tang from the raw sewage and rotting food scraps in the alley outside,"[26] as Nolen puts it. Agnes knows she is an important scientific specimen for researchers. Some, like Plummer, have built international reputations from studying women like her, with funding of tens of millions of dollars from research grants. Yet, as Nolen asks, "what does this project owe Agnes?"[27]

Agnes and women like her do get help from the researchers: treatment for a lot of medical conditions and counselling for safer sexual behaviour, as well as free condoms and referrals. Since 2004, she has been eligible for ARV treatment. But what about a way out—training for a new job? This would require her developing skills with numbers and in small-business practices. But "you can't get a research grant for that,"[28] Plummer told Nolen. "The ethics of science today," Nolen concluded, "require that the women get counselling and condoms, but ethics approval boards make no demands about math classes or instruction in how to set up an alleyway beauty salon."[29]

It's clear that the longer you work with a community, the deeper your responsibility becomes. That responsibility will become not just medical but social. Yet you also can't expect health researchers to be a development agency. This would undermine anyone doing research and make research on important topics like HIV impossible. Did Plummer have an obligation to get women like Agnes out of sex work? Should Mlisana try to persuade the teenagers in Vulindlela to stop trading sex for rides to town, or intervene with the menfolk to stop physical abuse? We think this would be an unrealistic expectation, although it's a laudable goal. To expect researchers to single-handedly reverse the years of injustice oppressing these communities is to expect too much of the research community. In our view, the obligation of researchers should be to recognize these background conditions of social injustice, raise awareness and work with others against these conditions, and ultimately leave a community less oppressed than it was when they arrived.

We think both Plummer and the CAPRISA researchers in Vulindlela are doing it right. In fact, we think Vulindlela is a model for researchers everywhere. In Vulindlela, researchers like Koleka Mlisana have acquired a full and intimate understanding of how the community works. They listened to the women and adapted the research trial to fit the women's lives. It's a great example of a research partnership between scientists and community, a partnership built on an understanding that respects cultural differences and practices. This deep community engagement takes tremendous effort, but we believe it's the only way to do a clinical trial successfully, especially if it's controversial.

We've long known in both the developed and the developing worlds that community support is crucial if you want to make sure the general public adopts a scientific innovation. It's crucial because a major obstacle, as we learned in interviews with scientific and medical leaders in South Africa, is the myths people have about new technology. The battle against HIV in South Africa is a perfect example.

William Makgoba, the Oxford-trained vice-chancellor of the University of KwaZulu-Natal, has seen the suspicion of new technology for himself. As former head of the South African Medical Research Council, he fought with South African president Thabo Mbeki over the government's laggardly policy on antiretroviral drugs for HIV.

"The cultures of Africa are in general very conservative about new things, about adopting new things," he said. "They're not only conservative, but they are very slow. They're not only very slow, but they also require champions that they can trust."

Makgoba told us for our 2007 study that the spokespeople for prevention of HIV have been doing a good job of relaying a clear and consistent message about abstinence and being faithful. But the leaders who have been championing the value of ARV drugs to save the lives of people who become infected with HIV have not managed to transmit such a clear message to the public. "The drugs are portrayed as a problem: they are poisonous, they will kill you and they are not the only thing to use," he said. The result: "A significant number is not actually interested in taking drugs, even pregnant mothers who are HIV-positive."

We believe that, given the right tools and involvement, the communities that are testing new technologies, such as microbicidal gels, can help. If these people deeply understand the new technology, they can become the researchers' most powerful allies when it's time to distribute the new technology to the general public. They can talk with people in other communities about the technology—its benefits, and its side effects, if any. They may be in a better position to demolish myths than the scientific and political leaders could ever be.

In CAPRISA's case, the approach of engaging the community, and also sticking to the approach of applying microbicide before sex, has already paid off. On July 20, 2010, at the International AIDS Conference in Vienna, the husband-and-wife research team of Quarraisha and Salim Abdool Karim announced the results of the clinical trial of tenofovir microbicide gel applied before sex: it reduced the risk of women becoming infected with HIV by 39 percent. CAPRISA estimated that widespread use of the gel "could prevent more than half a million HIV infections in South Africa alone over the coming decade."[30]

The approach of engaging communities may eventually be of help with the most controversial scientific trials—which, we soon discovered, are not the ones that involve sex. No, the most controversial research concerns the innovations that modify DNA, especially the DNA of food and bugs. How can researchers handle the public's suspicion about altering the code of life so that research trials can be conducted on potentially life-saving innovations?

The suspicion about genetically modified food puzzles many scientists. There seems to be a double standard. Who's complaining about recombinant drugs and vaccines, which are produced by genetic engineering? Some of the same people who shun GM food may even be taking advantage of our scientific ability to genetically engineer drugs if they manage their diabetes with shots of insulin. Almost all the insulin that diabetics take today is genetically engineered;[31] in fact, recombinant insulin was the first genetically engineered drug to be approved for therapy in humans.[32] Yet when GM techniques are used on food, it's a different story, as researchers from Switzerland and Germany discovered a decade ago.

Around the time we started our journey in global health in 2000, researchers reported some astounding news in the battle against vitamin A deficiency, which kills 6,000 poor people a day by weakening their immune systems,[33] and makes 250,000 people a year go blind.[34] In many parts of the developing world, millions of children, especially those under age five, begin to go blind in the evening. As night falls, they can see only vague shapes. They cannot play or study. Most of the children simply accept night blindness as the norm. Yet the cause is known: a deficiency of vitamin A.[35] The staples they eat every day—often just one food, like cassava or sorghum or rice—lack this important vitamin. It can be easy to fix. Just a balanced diet, or a drop or two of vitamin A in a capsule taken by mouth, will restore normal sight almost immediately. Yet the capsule intervention requires an army of public health workers, and many countries can't afford that.

Then, in 2000, two researchers reported[36] some astounding news. Professor Ingo Potrykus, a plant scientist at the world-renowned Swiss Federal Institute of Technology in Zurich, and Peter Beyer, a professor of cell biology at the University of Freiburg, showed that rice, a staple many children eat every day, could be genetically altered to boost its vitamin A content. This would prevent blindness and death. The two scientists inserted one gene from a common soil bacterium and two genes from daffodils into the rice so that beta carotene, the precursor of vitamin A, accumulates in the part of the rice that we eat. This genetic tweak turned the rice a golden yellow. (The term "Golden Rice" was proposed by a social entrepreneur in Bangkok, and the name stuck.) This radical new mode of delivery could alleviate vitamin A deficiency at a fraction of the cost of vitamin A supplements. And the inventors intended it to be free of charge.

Getting this marvellous invention to the villages and homes where it could save lives was an entirely different challenge, though. For starters, practically every step in the creation of the new rice, every gene, every laboratory process, appeared to be "owned" by someone or some company. There were altogether seventy intellectual property rights that belonged to thirty-two entities.[37] The researchers would need to negotiate with each one and perhaps pay royalties to many of them. If one or more of them refused, it could ruin their humanitarian effort. Yet this intellectual property issue was surprisingly quick to resolve. The inventors found a powerful partner in the agrichemicals multinational company Syngenta. The company suggested that, at most, only a few of the patents may have been infringed in the making of Golden Rice. In just a few months, it persuaded other multinationals to allow the humanitarian Golden Rice free licences for the project. The inventors would be able to donate the seeds to farmers in the developing world, for free, with no strings attached. It seemed an incredible gift to the poor world. What a great example of science serving the poor—at last. Potrykus was even celebrated on the cover of *Time* magazine in July 2000 under the headline, "This Rice Could Save a Million Kids a Year."[38]

Yet as of this writing, Golden Rice is still not in use, and millions of children are still going blind or dying. When we called Beyer to find out why, he sounded weary and frustrated. Golden Rice had stumbled into a towering roadblock—politics. Under the terms of the deal with the Golden Rice licensees, countries needed to have the necessary biosafety regulations to permit Golden Rice to be grown in the fields. Yet many countries, including those with high rice consumption plus a serious vitamin A deficiency, did not have such policies in place. Without them, they could not plant Golden Rice. And when the countries did develop regulations for genetically modified organisms, they were sometimes "unnecessarily burdensome" and "overly politicized," according to an article Potrykus co-wrote in the *Handbook of Best Practices on Intellectual Property*.[39] Writing in *Nature* in July 2010, Potrykus argued[40] that "regulation must be revolutionized" because "unjustified and impractical legal requirements are stopping genetically engineered crops from saving millions from starvation and malnutrition."

Helping to delay the approvals was the powerful anti-GM movement in Europe. Beyer could see the evidence of it every day as he was driving to work—a huge anti-GM sign erected by the activists.

European activists, on the warpath against Monsanto's genetically modified soybeans and other products, aimed their fire at Golden Rice. Beyer couldn't understand why. The researchers and companies weren't trying to monopolize a crop. This was a humanitarian project, with no commercial purpose. Yet the anti-GM lobby and their scientific allies didn't see it that way, and some still don't.

Although the World Health Organization[41] and a host of other significant panels, such as the GM Science Review in the United Kingdom,[42] say that GM food is safe, activists disagree. In early 2009, for example, thirty international scientists circulated a letter[43] sharply criticizing the testing of Golden Rice on adults and children. "Extensive medical literature" shows that retinoids that can be derived from beta carotene are toxic and cause birth defects, they charged. "In these circumstances the use of human subjects (including children who are already suffering illness as a result of vitamin A deficiency) for GM feeding experiments is completely unacceptable."

And yet independent ethical review panels in the United States and China, where the work was conducted, signed off on the necessary research. As well, supporters of Golden Rice counter that there is no credible evidence that any genetically modified crop is unsafe. After all, nature and plant breeders modify genes all the time. And the amount of beta carotene fed to each child in the trial was less than that in a small carrot. According to the researchers, the anti-GM mood in Europe has stopped the scientists from getting funding from European sources. Golden Rice is now sponsored by American funders such as the Rockefeller Foundation and USAID. "I am living on the wrong continent for this kind of research," said Beyer. "The attitude here is spoiled, completely irrational and stupid." Adrian Dubock, who has spearheaded the Golden Rice project for Syngenta for the past ten years, is equally frustrated: "Since I started, more than nine million people have died of vitamin A deficiency. That's more than died in the Holocaust."

Beyer and Dubock have every right to be angry. Here's a bunch of rich people from Europe telling poor people what they can't do to save their own lives. Critics typically point to the "precautionary principle," which states that if an action or policy might cause severe or irreversible harm to the public or to the environment, in the absence of a scientific consensus that harm would not ensue, the burden of proof falls on those who would advocate taking the action.

Critics have also raised some specific concerns that do need to be taken into account. Among these is the charge that Golden Rice does not contain enough vitamin A, and an individual would have to consume abnormal amounts of rice to get the necessary daily requirements.[44] This was a fair criticism in the early days, but the newer versions of Golden Rice have much higher concentrations: a very modest and achievable amount of just one cup of cooked Golden Rice-2 provides up to 60 percent of the adult recommended daily allowance of vitamin A. Another concern has been that the beta carotene in Golden Rice may not be stable or stably absorbed, but a recent study[45] concludes that it is indeed stable and stably absorbed into the body. In the body, the conversion rate to the active vitamin A is much better than for bright-coloured vegetables such as spinach and carrots.

Other critics view Golden Rice as a Trojan horse that will lead to more widespread use of genetically modified crops.[46] We understand these concerns, but think that individual crops should be judged individually, and their risks (including for health and the environment) and benefits assessed rationally and scientifically. What we do know is that genetically modified crops are now cultivated in more than twenty countries, covering more than a hundred million hectares, by fourteen million farmers, of whom thirteen million are resource-poor farmers in developing countries, mainly India and China.[47] Still other critics point to the possibility of underlying poverty and loss of biodiversity being compounded by corporate control. But Golden Rice will be offered at little cost and in a manner that is not under the control of corporations. Of course we should aim to provide varied nutrition-rich foods whenever possible, and we should work hard to ensure that, but it would be wise, where lives are concerned, not to let the ideal be the enemy of the good. We must also think about providing more than one solution. After all, if other theoretical and real solutions were adequate (for example, providing supplements, fortifying existing foods, teaching people to grow crops that contain vitamin A), we wouldn't have vitamin A deficiency of such magnitude in the first place.

A cautionary tale comes from events in Zambia during its famine in 2002. In the midst of that famine, the people in one village were so hungry that they stormed the chief's palace and made off with two thousand bags of corn. A week later, Zambia's president, Levy Mwanawasa, rejected a shipment of genetically modified corn from the United States. "Simply because my people are hungry," he announced, "that is no justification to give them poison, to give them food that is intrinsically dangerous to their health."[48] His hungry people didn't feel that way. Shortly after the president's announcement, government officials summoned aid workers in the northwestern part of the country to demand why they were still distributing GM corn to 125,000 refugees in five camps. Aid workers replied that riots would break out if they withheld the food.[49]

In the words of Hassan Adamu, Nigeria's former agriculture minister, "To deny desperate, hungry people the means to control their lives by presuming to know what is best for them is not only paternalistic but morally wrong."[50]

Beyer hasn't given up. He's now leading an international effort called the ProVitaMinRice Consortium, funded by the Gates Foundation, to develop new varieties of rice in addition to Golden Rice that will boost the quality and content of protein and include iron and zinc, which are also important to a healthful diet. The first introduction of Golden Rice is expected to be in the Philippines, where up to 55 percent of children under five suffer from vitamin A deficiency despite a capsule program that has been in place since 1993. But before this can happen, the multinational and medical communities, politicians, religious leaders, farmers and consumers all need to be convinced of the benefits of Golden Rice. Science is only a small, although very necessary, part of behaviour change needed for public good.

We think the South African community of Vulindlela could offer a good model for involving people in the acceptance of GM food. You start locally, with one or a handful of communities. You get to know the people, understand their problems, and deal with them appropriately. You quietly build up trust between researchers and the community. Then they will be your strongest advocates. When other communities see the children in the Golden Rice testing communities growing up without going blind, they'll want to try it too. They won't be unduly swayed by the concerns of European activists who may not necessarily make their first priority the health of people in the poor world.

The Golden Rice story is an important lesson for other scientists who want to modify the genetics of food to improve nutrition and save lives. As we saw in Chapter Two, four teams in the Gates Foundation Grand Challenges, including Beyer's, are making amazing progress in the mission to insert vital nutrients—iron, vitamin A, zinc and in some cases proteins—into staple crops. In recent years, they've used genetic engineering to insert these nutrients in bananas, cassava, rice and sorghum. The testing has been confined mostly to the lab, but now they're moving towards trials in greenhouses and even in fields under controlled conditions. How will they confront the challenges that have hindered the Golden Rice group for the past decade?

To answer this question, we talked to Dr. Florence Wambugu, the Kenyan crop scientist who persuaded fellow scientific luminaries to endorse GM foods as one of the grand challenges funded by the Gates Foundation, and who is a lead researcher in one of the nutritionally enhanced crop projects that arose from this discussion. She's now working on a project to biofortify sorghum, a grain that is the staple for millions of Africans.

Wambugu grew up on a farm where her family made a meagre living by cultivating sweet potatoes. Working with her hands, she could see for herself how many sweet potatoes were lost to diseases and pests. She also worked hard at school, and her mother eventually sent her to a girls' boarding school with proceeds from the sale of the family's only cow. She went on to become one of the first women in Kenya to attend university, and later obtained her doctorate from the University of Bath in England. Her thesis was, no surprise, on the sweet potato. After a stint in the United States, where she worked for Monsanto for three years, Wambugu worked on a GM sweet potato that is resistant to viruses. It was the first transgenic crop developed for Africa. Early results from field trials in 2001–2 were disappointing, but the experiments led to a biotechnology policy system that set the stage for further research in GM crops.

Now, as founder and CEO of the non-profit Africa Harvest Biotech Foundation International, Wambugu is trying to genetically improve sorghum, to improve the digestibility and quality of the protein and add higher levels of vitamin A, iron and zinc. The result is known as Biosorghum, or Africa Biofortified Sorghum (ABS), and it's moving towards confined field trials in several African countries.

Progress hasn't been easy. When Wambugu's foundation applied to a South African regulator to test fortified sorghum in a confined greenhouse, activists claimed that Biosorghum was a "wholesale contamination of Africa's prized sorghum heritage." The government sided with the activists.

Fortunately, that didn't end the Biosorghum story. By 2008, the South African government had reversed its position and issued a permit for the greenhouse trial. ABS is also on its way towards testing in greenhouses and field trials in Kenya and Burkina Faso.

Wambugu is slowly making headway, despite opposition from a powerful anti–GM food campaign, because she's reaching out to the public. "You have to somehow get the information through the radio, newspapers and the TV, but sometimes, to get to the grassroots you have to go to the community through schools and churches. We have to get to the people," she said. "We have to explain that the GM sorghum will enhance their local varieties with better nutrition and yet the farmers can keep their own conventional seed. No matter how much misinformation, lies and tricks, we counteract lies with the truth. We support the integrity of science. The truth stands."

Wambugu will tell you something else: if researchers want to develop effective products for the developing world, they'd better collaborate at a research and leadership level with scientists who come from that world. Policy-makers in every country tend to trust their own experts, and the developing world is no exception.

But in the end, it will not be scientists like Wambugu who dispel the myths about GM food. It will be the neighbour, the chief, the nurse, the teacher. Once they embrace the idea that biofortifying food can save their lives and their communities, they'll create a powerful force for change, which will impress regulators and farmers alike.

That leaves one, even bigger hurdle. Communities may ultimately accept GM food, but what will they think of genetically modified insects, such as mosquitoes?

Mosquitoes are the ubiquitous distribution system for some of this planet's worst killers, such as malaria, and they are maddeningly hard to attack. They can slip through bed nets, avoid insecticide spray or become impervious to insecticide. The damage mosquitoes cause in humans is devastating. The *Anopheles* mosquito distributes the *Plasmodium* parasite, which infects up to a quarter of a billion people a year and kills almost a million children under five in sub-Saharan Africa; indeed, one child dies of malaria every thirty seconds.[51] The *Aedes* mosquito spreads a virus that causes dengue fever in fifty million people a year,[52] mostly in the Southern Hemisphere. The virus gives its victims a high fever, headache, swollen glands, painful red eyes and pain in the joints, muscles and back. It can get a lot worse: up to half a million people each year who get infected a second time develop dengue hemorrhagic fever, in which the virus causes blood vessels to swell and leak into tissues, producing purple bruises and black vomit. In the worst cases—an estimated twenty thousand a year—the victim suffers an ugly, brutal death. It's especially scary because there is no drug to treat this disease, and no vaccine to prevent it.

Scientists have tried to disarm the mosquito once before. In the 1970s, the World Health Organization and the Indian Council of Medical Research hatched a plan to control dengue fever by sterilizing male mosquitoes through irradiation (a method that had successfully eradicated the parasitic New World screwworm fly in the United States a few years earlier). But then a parliamentary committee claimed that the true goal of the research was not disease control but rather biological warfare—to spread yellow fever. Amid the ensuing furor, the trial was quashed.[53]

A human generation later, the mosquito is once again the focus of several high-powered research projects around the world that share the same aim—to make it impossible for the insect to spread malaria or dengue fever. This time, researchers have an infinitely more powerful tool than irradiation. It is the ability to change the genetic recipe of the mosquito to prevent it from spreading disease. They have many clever and exotic strategies. They can shorten the life of the mosquito so that it dies before delivering the virus or the parasite. They can make the mosquito an inhospitable host for the parasite. They can give the mosquito a new power, to poison a parasite inside its gut. They can cripple or kill the female mosquitoes that spread disease. Whatever the strategy, they all rely on the new ability to rewrite bits of the mosquito genome. These are GM mosquitoes. How can scientists persuade the world to test them?

One of the brightest stars in the mosquito world today is Anthony James, professor of microbiology and molecular genetics at the University of California, Irvine. He's an open, gregarious scientist who has been publishing thrilling articles about the manipulation of what he calls "the most dangerous animals in the world" in the world's primary scientific journals. James is also well equipped for the voyage from lab to village. He grew up in the racially charged Los Angeles of the 1970s, and as a child of a mixed marriage, he was highly tuned to the politics of civil rights. "I grew up as a black man," he said when we talked to him in the fall of 2009, "but now I think I'm African American, like our president."

His own history has instilled in James a social conscience that makes him reject a paternalistic way of doing things. "Principle number one," he said to us, "and the single biggest lesson, is to talk to people early. The biggest mistake is to show up with a finished product and say, 'Use this because I say it's good for you.' Once you've lost the attention of the community, it's difficult to rebuild."

James knew he would need to do things differently as he started to modify the mosquito in his lab in Irvine, California. When we spoke with him, he was preparing to test some of his genetically modified mosquitoes in a small town in Mexico, and he was bubbling with excitement about the incredible scientific challenges he faces every day. He made the task of intercepting the parasite on its journey inside the mosquito sound like a prime-time thriller.

When the female mosquito bites a person with malaria, he explained, it picks up some parasites along with the blood meal. The one-cell parasite has to move into the mosquito's midgut, a potentially hostile environment with a different temperature and acidity than the warm bath of the human host. There it develops and grows and then makes its way to the mosquito's salivary gland, where it stays until the mosquito bites again and deposits it into the next human. The parasite, of course, doesn't have eyes. The receptors on its surface tell it where to go. This gives scientists several opportunities to interfere with the parasite's journey inside the mosquito. One tactic is to change a gene the parasite needs to find its way to the midgut. Without that gene, the parasite gets lost.

James's favourite strategy, though, is a wild one. It's based on the fact that malaria parasites are highly focused killers. A parasite that causes malaria in humans, for instance, won't cause malaria in mice. This is because the mouse's immune system can disable the parasite with an antibody that covers it like a sheet. James and his lab mates cloned the parts of the genes that make the cloaking device. Then he put the mouse's cloaking gene into the genome of the mosquito. James was so excited by the idea that he sounded like a kid with a cool new computer game: "Ain't that the dangest? We're giving the mosquito a piece of the mouse immune system that makes the mouse resistant to human parasites. It's a true chimera."

The idea is to give the GM mosquito the artillery—borrowed from a mouse—to disable the parasites inside its gut. The parasites will never make it to the salivary gland, so the next time the mosquito bites a human, it won't deposit the parasites that cause malaria. To see if the model works, James did an experiment using the *Aedes* mosquito, the one that spreads dengue viruses and a malaria species that infects chickens. It was a success: the mosquito with bits of the mouse immune system was able to stop the parasite that infects chickens. James's group is now working on several experiments to see if this strategy works on malaria parasites that infect humans.

There is a far bigger challenge, though. Even if you can change the DNA of a few mosquitoes in the lab, how do you alter all the mosquitoes in a town or a city or a continent so that they don't spread disease? Finding the right gene drive system, as it's called, is critical. "It's fairly complex," said James. One key challenge is to make GM mosquitoes strong enough to outlast and outmuscle the wild ones as propagators. You have to engineer a mosquito that not only has the ability to disarm the parasite but also is strong enough to eventually take over the whole population.

James is making considerable headway with dengue fever, too. The disease is transmitted by the female. James and his colleagues wondered if they could stop the mosquito from flying, so it could no longer spread the disease.

They discovered an intriguing way to do it. When the female mosquito takes a blood meal, she quadruples her weight and therefore needs strong muscles in her wings to fly. James and his team discovered the gene that turns on when the muscles in the female mosquito's wings are growing. His collaborators at Oxitec, the biotech company that is testing GM malaria mosquitoes, deprived the females of their flying muscles. "We've got these great videos of them lying around the cage," James said. "They're just sitting there waiting to be eaten."

How do you ground a whole population of female *Aedes* mosquitoes? It turns out that you don't need to release that many GM mosquitoes to take over. In a small town, it may only take five thousand GM mosquitoes—a couple of containers of them—to alter the genome of the *Aedes* population. For a city of half a million people, James said, "we think we could easily do it with five million mosquitoes a week." In other words, over six to nine months, you'd have to release a total of 200 million mosquitoes. "It sounds like a lot but it's not." It would take about twenty mosquito generations to permanently alter the population of mosquitoes in a region. And it's better than insecticide, James said: "We're using the innate drive of the males to find the females."

Safety is an obvious concern. "We continually try to categorize all the possible ways this might screw up the environment," said James. What would happen if a certain species of mosquito disappeared from a region? James thinks the answer is nothing. There's no evidence that mosquitoes are a keystone species that are vital for other organisms to survive. Indeed, in most places mosquitoes are invasive, non-native species that live in niches created by human activity. What would happen if the gene that has been put in the mosquito got out? "One of the good parts of molecular genetics is that we can design features so they don't work in another species," said James. "We can build in exclusivity." People often say to James: 'But you don't know what's going to happen, do you? How do you develop a plan for the unimaginable?' There's always insecticide. But if a swarm of GM mosquitoes starts to cause trouble, James says scientists do have a plan. They can release a new group of mosquitoes that are genetically designed to search out and attack the bad ones.

While plenty of research is taking place in labs in Europe and the United States to tweak the mosquito so it won't spread malaria, James isn't ready yet to test his GM versions in the field or on humans. "Malaria is extremely complex," he said.

But he is preparing to test out, in giant cages in Mexico, a GM mosquito that won't spread the dengue virus. He would not be the first to test GM mosquitoes outside the confines of the laboratory. Luke Alphey, a former senior research fellow in the Department of Zoology at Oxford, founded Oxitec to control pests through what the company calls "birth control for insects." They use the same technique that worked so successfully to eradicate the New World screwworm in the United States. The company releases millions of sterile insects over a wide area to mate with the natives. They produce nonviable eggs; the insect population declines. But instead of sterilizing the insects through irradiation, as was done with the screwworm, the company uses genetic engineering (on males only; sterile females cause too many problems). The GM males are fit but sterile—and safe, asserts Oxitec. Even if the insects were to escape from the cages where they're reared, they can't produce offspring.

These GM mosquitoes were tested in confined semi-field trials in Kuala Lumpur in 2007–8 by the Institute for Medical Research of the Malaysian Ministry of Health. The tests constitute the most realistic trials of engineered mosquitoes to date. Now Oxitec is working with the Malaysian government, the Institute for Medical Research and regulatory/ethics committees to get permission for a limited-release experiment in an isolated ecosystem.

Already, the anti-GM drums are sounding. The first objection came from the watchdog group ETC, which is the same group that gave the name "Terminator seed" to a particular Monsanto product.[54] In January 2008, Jim Thomas of ETC expressed serious concern to Wired. com[55] about any mosquito releases: "Releasing millions of genetically modified terminator mosquitoes into wild ecosystems amounts to a reckless and uncontrolled experiment with a risky technology. Oxitec's [project] abandons all pretense of containment or possible recall. I wonder what sort of liability they are willing to assume if something goes wrong?" An online petition[56]—"No to GM Mosquitoes Release in Malaysia"—was sent to the Malaysian Ministry of Health; it called for public consultation and independent scientific evaluations before releasing the GM mosquitoes in Malaysia. "We ask that organizations (whether private or governmental) involved in the eventual release of the GM mosquitoes into our Malaysian environment take full responsibility and must be liable for any problems that may arise from the release of this biological control."

If the opposition to GM food is any indication, this reaction to GM mosquitoes could spell trouble for the Oxitec group and for Tony James in his fight against the dengue virus. Is there a way to get these potentially life-saving innovations to the villages where they're needed without hitting a wall of opposition? We think so.

In 2005, when we met Tony James at the Seattle meeting that kicked off the Gates Grand Challenges, we were accompanied by Jim Lavery, who did his PhD with Peter and is now one of the world's leading experts on the ethics of international research. He was the first bioethicist to work at the John E. Fogarty International Center, the international division of the U.S. National Institutes of Health. Now he works at St. Michael's Hospital in Toronto and co-leads our grant to provide ethics advice and support to researchers in the Grand Challenges for Global Health.

That day in Seattle, Tony James and Jim Lavery sat down together in an unusual setting—an old dinner theatre—and discussed how to run a trial of GM mosquitoes in an ethical way. James had the money to fund a trial of his GM mosquitoes, but he didn't know how to select a community for the research. "We don't know how to do this either," said Lavery, "but I have a feeling we are going to have to start working together."

James agreed. "So how do we select the site to do these field trials?" he asked. James had already narrowed down the options: he was determined to test the GM mosquitoes in the South. If the GM mosquitoes worked as planned, they were going to be used in Southern countries anyway, so why test them in the North just to say, "If they're good enough for me, they're good enough for you?" Wouldn't that be paternalistic? There were a few options: Thailand, Peru, Mexico and Trinidad and Tobago. The countries were all fine from a scientific point of view, but James needed a site where the researchers could satisfy regulatory requirements. And most important, he had to choose a site where he could work closely with the community. For both men, this was a step into unknown territory, and James wanted to get it right. "I don't want, five years from now, someone to say, 'Look at this guy, he didn't care.' If this project has longevity, it will have longevity both because of the science and because of this community dimension."

In March 2007, Lavery flew to Trinidad to see whether it might be a suitable test site. He soon spotted a couple of problems. First, there was no regulatory regimen in place to allow GM mosquitoes, and although a draft bill was before the parliament, it wasn't clear whether it would be approved. Second, the roads on this beautiful tropical island were too narrow to handle big trucks carrying equipment to build cages the size of warehouses. The villages were tightly packed too, which would mean that people would be living right next door to the mosquito cage and might not like it. He had reason to be concerned—a young sociologist on the island had done some preliminary public opinion polling about GM organisms and had found some pockets of resistance. This, for Lavery, was a big red flag. "We thought Trinidad could be workable," he said later, "but there were some important hurdles that would need to be cleared. I was extremely impressed by the local collaborators, but the fact that the bill was hanging in parliament made me a little nervous."

A month later, Lavery flew to Mexico and headed for Tapachula, a small city close to the Guatemalan border in the southern state of Chiapas. This town is proud of its role in the fly business—and in combatting the Mediterranean fruit fly. This pest, medfly for short, had caused many millions of dollars of damage to fruit crops in California, Texas and Florida in the 1980s. Massive spraying (ordered by, among others, California governor Jerry Brown, a strong environmentalist) helped, but the final stage of the campaign was to release millions of sterile male medflies to stop the population from reproducing. These sterile medflies are still being used, and many of them come from the medfly facility at Tapachula. This facility, funded in part by the U.S. Department of Agriculture, uses radiation to sterilize one million Mediterranean fruit flies every day to ship to the United States. It's a successful program that has helped to eliminate medflies in North America and all the way down to Guatemala. For the people of Tapachula, it's a Mexican success story, a matter of local pride.[57]

Vector control—the methods of controlling insects such as mosquitoes that transmit disease in a community—is not something new in Tapachula. The Mexican Ministry of Health Malaria Research Centre is located there, and one day Lavery travelled with the vector control officials and Tony James's collaborating entomologists to the outskirts of the large town to hunt for the *Aedes aegypti*, the mosquito that transmits dengue, which likes to live indoors. The officials, dressed in public health vests with a little insignia on the shoulder, stepped into the small brick houses to inspect them. Using pipettes, they'd catch mosquitoes from behind the curtains or from under the bed. The homeowners, often women carrying babies on their hips, were generally willing to grant the officials permission to enter their homes.

Lavery was impressed. "The vector control teams have a presence in the community already and people seem to understand their role and accept them in the way communities in the North might accept people coming to their homes to read their water meters."

The threat of dengue is not theoretical in this community: the children here are affected and sometimes killed. The community has a municipal dengue control committee, made up of nurses from the local health clinics as well as business leaders and other members of the community. The committee's mandate is to study the dengue problem and think of what to do to protect people, and then report to the mayor. Lavery, along with Janine Ramsey, then the head of the vector control research centre in Tapachula, met with the committee in Ramsey's office. By the end of the meeting, it was clear to Lavery that the committee had no reservations about GM mosquitoes, probably because of their experience with the nearby medfly facility. They wanted to complement their existing arsenal of mosquito-control measures, which are never completely successful. Mexico had another thing going for it—legislation allowing the GM mosquitoes. That tipped the scales. Tapachula was chosen as the site to test James's GM mosquitoes.

James hopes to start testing the GM mosquitoes in cages in the town soon. Community engagement, he knows. is a critical part of the job, which is why he hired an anthropologist originally from Tapachula to help. "We need an example of success with GM organisms," he said. "The champions have to be the end users, rather than the suppliers. If you give the emotional and intellectual ownership to people who are going to use it, it becomes theirs. You're giving them the ability to say yes or no."

This vital part of the exercise—spending time with the community in which you've chosen to test an innovation—may seem to be outside the conventional scope of scientific concern. Indeed, that is precisely what many scientists tell James. "It astounds me how many people just don't get it," James said. "My response is: it's the right thing to do."

We couldn't agree more.

CHAPTER FIVE

In this chapter, we will tackle head-on one of the biggest, most contentious and perhaps the trickiest of all the roadblocks that lie between the lab and the village: the urgent need to engage multinational corporations and businesses as partners with the public sector in promoting health in the developing world. The obstacle is, of course, a profound distrust of multinational corporations on the part of governments, academics, activists and health workers. To many people who work in global health, these corporations and businesses are villains. Certainly, they need to be regulated, monitored and prevented from doing harm. But many global health workers also think these businesses shouldn't be engaged at all as partners in promoting health in the developing world.

This distrust is perfectly understandable, given the past behaviour of some multinational corporations in poor countries. In the 1970s, Nestlé peddled baby-milk substitutes to low-income mothers, which led to sickness and even death in babies in many poor countries.[1] After it was hit by an international boycott in 1977, the company vowed to end the practice, and yet a 1997 study,[2] "Cracking the Code," showed that Nestlé and other multinational producers continued to hand out samples of infant formula to poor mothers in several parts of the developing world.[3] Today, the companies claim they have ended these practices, while various groups claim they have not.[4]

Pharmaceutical companies, for their part, hit a low point in the mid-1990s as HIV was decimating South Africa. Since most people there couldn't afford powerful but expensive antiretroviral drugs to save their lives, Nelson Mandela's South African government proposed legislation that would allow the importation or local manufacture of cheap copies of the life-saving drugs. That's when GlaxoSmithKline and several other pharmaceutical multinationals struck back with legal action against the government, to protect their patents. The legal move caused an international uproar against the multinationals.[5]

Meanwhile, Monsanto, the leading creator of GM seeds, was accused of trying to take over the world's crops. Suspicion of Monsanto's motives was so great that, as we saw in the previous chapter, the president of Zambia refused to distribute GM corn in his country, even while millions of people were starving.[6]

These scandals have created a climate of distrust that is unfortunately blocking significant progress in current efforts to relieve one of the biggest problems in global health—hunger and malnutrition. Malnutrition is implicated in the deaths of between three and four million children under the age of five in developing countries every year.[7] Even when it does not kill, malnutrition is stunting the physical and mental growth of hundreds of millions of people in the developing world.[8] In this chapter we will show how, in order to change this terrible picture, we need to engage multinationals. They hold the patents to innovations and have the know-how to develop and distribute innovations to people who need them in the developing world. Many large companies are showing, through their actions, their willingness to help developing nations with advances in health, nutrition and agriculture. Some senior executives, such as Derek Yach at PepsiCo, are themselves global health activists who have moved into the private sector because they think they can achieve more inside powerful global corporations than they can inside international organizations, academia, governments or NGOs.

We need to engage multinationals because it is clear now that the public sector alone cannot solve the profound health problems of the developing world. The public sector simply doesn't have all the resources or the expertise needed. Those who work on these issues within the public sector can succeed only if they collaborate with business. When and where it has been allowed to happen, this kind of collaboration, through public-private partnerships, has been tremendously successful in developing drugs, vaccines and diagnostics for the diseases of the poor. To give just one example, precisely this kind of partnership has created real hope for an effective vaccine against malaria within the next few years.

Yet private and public players have been unable to find a way to collaborate to save millions of young children from physical and mental stunting, and sometimes death, as a result of malnutrition. To state the problem baldly, the two sides are unable to collaborate because they don't trust each other. Many politicians and health activists expect the worst from multinationals, while many business leaders are reluctant to join a humanitarian project that has no guarantee of large profits or could expose them to a potential backlash and tarnish their brand in rich countries. And indeed, it is an immense challenge to establish trust between, for example, breast-feeding advocates and food companies. But we believe this trust must be built, and we contend it can be built, step by step. It is not an easy task, we know, but it's an absolutely crucial one for all of us in global health who want to save lives. And as we'll see, this process of building trust is already beginning to happen, from the ground up.

Today, a few of the same multinationals that were so sharply criticized in the past are making genuine efforts to improve health in the developing world. GlaxoSmithKline is working on one of the promising new vaccines for malaria; other multinationals are genetically modifying crops to make them tolerate droughts and resist pests; and other businesses are fortifying foods with vitamins to help children grow up healthy and resist potentially deadly infections. Equally important, some of these multinationals are finding ways to offer these innovations to people in poor countries at affordable prices. After decades of corporate indifference to the diseases of the poor, this is a welcome change.

It also gives us a very different picture from the one we saw when we began our journey into global health over a decade ago. In the pharmaceutical sector back then, for instance, the pipeline of new drugs and vaccines for diseases of the poor was almost dry. As researchers reported[9] in the medical journal The Lancet in 2002, a total of 1,393 new chemical entities were marketed between 1975 and 1999. Of these, only 16 were for tropical diseases and tuberculosis—the diseases of the poor. Some studies provided slightly higher figures,[10] but the overall picture was one of neglect. From the point of view of anyone who cares about global health, this was an abysmal record.

The problem was rooted in the logic of the market that prevailed at the time. The multinationals claimed that it cost hundreds of millions of dollars to discover, develop and bring to market a new drug, and therefore they had to seek ways to recoup that investment in the profits from sales of drugs. Their profits, however, were consistently among the highest compared to other industries. And in any case, people who were sick and dying in poor countries were often too poor to afford life-saving drugs at *any* price. Following this logic, multinationals, even those with good intentions, had no market incentive to develop remedies for the poor. The accepted wisdom was that it simply did not make business sense for them.

Some companies, such as Merck, donated drugs as an act of charity. In 1987, Merck decided to donate the drug Mectizan in perpetuity to eliminate river blindness, a debilitating disease caused by the parasite *Onchocerca volvulus* (hence its other name of onchocerciasis).[11] It is transmitted by flies that breed near fast-flowing rivers. As large proportions of exposed communities go blind, this disease has become a major public health problem in Africa and elsewhere. Merck had to overcome phenomenal obstacles[12] to get the pills to the millions of people who needed them. Health workers had to walk for days to reach some villages, and political unrest posed a very real danger along the way. In some places, they found that half of the inhabitants were blind, and yet there was no formal health care system to deliver the drugs to prevent more blindness. But by teaming up with existing NGOs that reached remote villages, and by working with government health officials to integrate the Mectizan program into the national health care system, Merck succeeded. According to the company, the program now reaches eighty million people each year to prevent river blindness in Africa, Latin America and the Middle East. Nearly one-third of the at-risk population in the Americas is no longer contracting the disease, including the entire country of Colombia.[13]

Merck is not the only pharmaceutical company to have donated drugs. Pfizer donated[14] the antibiotic azithromycin to control trachoma, a bacterial disease that is carried by domestic flies.[15] The infection scars and inverts the eyelid, causing the eyelashes to rub against the front of the eye, gradually scarring the cornea, which then becomes opaque to light. If left untreated, it may lead to complete blindness.[16] The Pfizer initiative, together with the World Health Organization's SAFE strategy (Surgery, Antibiotics, Facial hygiene and Environmental improvements), has contributed to reducing the incidence of trachoma by eighty-nine million cases from 1997 to 2008.[17] Another hopeful sign: lymphatic filariasis, known as elephantiasis, has been controlled in China, Sri Lanka and other endemic countries with drugs donated by GlaxoSmithKline and Merck.[18]

While the results in these few cases have been spectacular, charitable drug donation is not really a remedy that can be applied to all of global health, nor is it a dependable solution. No other similar programs have lasted so long, on such a large scale, and with such impressive results as the Mectizan program. Charity may even have unintended, less positive consequences, making poor governments dependent on rich countries and, in some cases, complacent.

Fortunately, there is another way to make desperately needed drugs and vaccines for the poor: by encouraging the discovery and development of new drugs and other health products through public-private partnerships. These not-for-profit PPPs, as they are known, emerged in the mid-1990s; they involved multinational companies making use of pharmaceutical knowledge in a more sustainable way that did not require charity by the private sector acting alone.[19] Public groups, such as the Rockefeller and Gates foundations, put up the money, while the private sector donated intellectual property and expertise. One way to envision these partnerships is as a global, virtual pharmaceutical company for diseases of the poor. The intent is for any successful drugs emerging from these PPPs to be made available to poor countries at an affordable price. Meanwhile, the companies won't end up with a drug or vaccine or diagnostic that hardly any of the potential beneficiaries can afford.

One of the earliest PPPs, the International AIDS Vaccine Initiative (IAVI), was created in 1996 following a meeting of HIV experts organized by the Rockefeller Foundation.[20] Headed by Seth Berkley, who came from the Rockefeller Foundation, IAVI focused on research into and development of HIV vaccines for the developing world, as well as on building the developing world's capacity to conduct clinical trials in places where future vaccines would be tested. It also acted as an advocate for HIV vaccines, and it built a consortium with many other partners to address the gaps in knowledge needed to make an effective vaccine.[21]

Many other global health–related PPPs followed, and they have started to produce results. These include the Drugs for Neglected Diseases initiative, which has developed a combination treatment for malaria;[22] the TB Alliance, which has three tuberculosis drug candidates in clinical development;[23] the Meningitis Vaccine Project, which developed a new vaccine for meningitis;[24] and the Foundation for Innovative New Diagnostics,[25] which developed a tuberculosis test that can both diagnose the disease and test whether it is resistant to the usual drugs used for treatment, in under two hours in the clinic without the need for a sophisticated laboratory. There are currently about thirty product-development PPPs worldwide, and collectively they have a robust pipeline of drug and vaccine candidates in various stages of development, all meant for diseases of the poor.

This is a significant and welcome improvement over the bleak situation of a decade ago.

One prime example of the success of such PPPs can be found in the battle against malaria we mentioned earlier. A quarter-century ago, not one malaria vaccine was in clinical trials. As of July 2009, there were nineteen vaccine candidates against malaria in clinical development.[26] One of the most exciting prospects is the RTS,S vaccine. It dates back to 1987, when GlaxoSmithKline Biologicals spearheaded its early development in close collaboration with the Walter Reed Army Institute of Research in the United States. In January 2001, GlaxoSmithKline and the PATH Malaria Vaccine Initiative, which is supported by the Gates Foundation, embarked on a mission to develop the vaccine for young children in sub-Saharan Africa.[27]

The RTS,S vaccine candidate is made possible by genetic engineering: it fuses a protein from the malaria parasite itself with part of the hepatitis B virus.[28] So, in addition to inducing partial protection against malaria, RTS,S is also designed to protect against hepatitis B.[29] The results have been encouraging. The vaccine underwent small-scale testing on adults in the United States and Belgium. Then, beginning in 2003, it was tested on more than two thousand children in Mozambique.[30] The result: it reduced clinical malaria by 35 percent and severe malaria by 49 percent over eighteen months. When it was tested on infants, the group most vulnerable to malaria, it reduced infection by 65 percent over three months of follow-up.[31] Although it's only partly effective, the RTS,S vaccine is a major breakthrough since no other safe vaccine with the potential for large-scale rollout has ever worked like this. At the time of writing, it is undergoing Phase 3 trials on a larger set of children in many African countries before it can be licensed. The plan is to test sixteen thousand children across sub-Saharan Africa.[32] As Melinda Gates told a powerful audience gathered at Davos in the winter of 2010, "We will have a malaria vaccine, absolutely, in our lifetime."[33] This is a tremendous achievement in a relatively short time—and it is possible precisely because of a committed public-private partnership.

Other PPPs—called delivery PPPs—are designed to take innovations that the partnerships themselves may or may not have been involved in developing and market and distribute them in poor countries. The prime goal of the Global Alliance for Vaccines and Immunisation (GAVI), for example, is to facilitate access to underused vaccines where they're needed in developing countries. It uses donor funds to finance immunization in poor countries, in such a way that it drives down the price of those vaccines. GAVI says that in its ten years, it has helped to prevent more than five million deaths.[34] Meanwhile, the Global Fund to Fight AIDS, Tuberculosis and Malaria has grown to be the main source of funding for the control of those diseases. Created in 2002, the Global Fund provides a quarter of all international financing for AIDS globally, two-thirds for tuberculosis and three-quarters for malaria, with approved funding of $19.3 billion for more than 570 programs in 144 countries. It is considered one of the most successful initiatives in global health and has made a big difference in many people's lives: it has provided 2.8 million people with antiretroviral treatment, helped provide funds for the treatment of seven million new cases of infectious tuberculosis and provided most of the funding for 122 million bed nets and 142.4 million malaria drug treatments.[35]

Clearly, the PPP movement has enjoyed key successes over the past decade—in part because the very premise, the structure and philanthropic funding have solidified trust among the partners. And so we wondered: could the private sector also help to address another gaping hole in global health—malnutrition of children? As we have discussed above, under-nutrition and malnutrition are massive problems in the developing world. A landmark 2008 series[36] in the medical journal *The Lancet* confirmed the grim facts, and pointed out that deficiencies of micronutrients are a huge problem. A lack of vitamin A, for instance, causes over 500,000 deaths a year in children, while deficiency of zinc causes 400,000 deaths. Insufficient breastfeeding, especially in the first six months of life, causes 1.4 million deaths every year.

The number of deaths is horrifying, but sadly, this is only one part of the picture. Even if they survive, children who are malnourished in their first two years are scarred for life. Not only will they be shorter, they will perform less well in school, earn less as adults and give birth to smaller children. Malnutrition not only stunts bodies, it permanently stunts the brains of hundreds of millions of children. Talk about committing an entire generation to poverty. What is so vexing about this problem is that many of the deaths, and the long-term consequences for those who live, are easily preventable.

Many of the problems can be avoided if babies are properly nourished in utero and in their first two years of life. Those years are critical. This is when babies need to get the key nutrition they require for healthy brains and bodies.[37] What's the best way to make sure babies in the developing world get the nutrients they need? The answer for the first six months of life is obvious: infants should be exclusively breastfed. Breastfeeding provides good nutrition, and it also passes along the mother's immunity through antibodies that help the baby fight infections. Baby formula is much worse for babies in the developing world than for infants in richer countries—not only does it lack antibodies but when prepared with dirty water, as is often the case in poorer countries, it can be a source of infection.

Breastfeeding initiated early saves lives, but the rate of exclusive breastfeeding during the first six months of life is far too low in some developing countries, especially as women engage in the workforce and move to urban centres. In Bangladesh, for example, only 43 percent of infants under six months are exclusively breastfed, while in fast-growing Vietnam the figure is only 17 percent.[38] The overall figure in the developing world is 37 percent.[39] These numbers need to be vigorously increased by the public sector.

After the first six months, young children need more than just mother's milk because breast milk lacks essential micronutrients, especially iron, zinc and vitamin A.[40] This is why the WHO recommends that babies aged six to twenty-four months should ingest nutrient-rich complementary foods in addition to breast milk.[41] In 2008, this strategy got a boost from a high-level group of economists who gathered in Copenhagen to analyze the cost and effectiveness of various interventions to tackle global problems, including health.[42] This group, which included five Nobel laureates, concluded that out of forty interventions, the most cost-effective contribution to global development was supplementing children's diets with vitamin A and zinc. "Providing micronutrients for 80 percent of the 140 million children who lack essential vitamins in the form of vitamin A capsules and a course of zinc supplements would cost just $60 million per year," the Copenhagen group noted.[43] That investment would yield enormous benefits—more than $1 billion a year—in the form of better health, fewer deaths and increased future earnings. These are powerful and sound arguments, and yet, a few years later, we're still living in a world where eight million children under five die every year,[44] with malnutrition contributing to one-third of these unfortunate and unnecessary deaths. Deaths from malnutrition in the developing world vastly outnumber those in the developed world. Eighty percent of the world's undernourished children live in sub-Saharan Africa and South Asia.[45]

The Copenhagen group's insight remains as important as ever: to address under-nutrition of children aged six months to two years, you have to improve the level of micronutrients they ingest. Increasingly, we're realizing that one of the most effective ways to do this is to fortify food with micronutrients. Fortified foods are familiar to those of us in rich countries. The milk in our refrigerators is fortified with vitamin A to prevent blindness and strengthen the immune system and with vitamin D to prevent rickets; our cereal is fortified with iron to prevent anemia; our salt is fortified with iodine to prevent thyroid goitre and mental retardation in young children; and our drinking water is fortified with fluoride to prevent tooth decay.

In poor countries, however, the majority of people rely on a single staple, such as rice in Asia or cassava and sorghum in Africa, which are poor in micronutrient content. They often don't have access to meat, which contains protein and vitamins, or to fresh fruits and vegetables containing vitamin C (to prevent scurvy) and many essential B vitamins (such as folate to prevent spina bifida in pregnancy and thiamine to prevent beriberi). This is why it's so important to fortify the limited diet that people in these countries do eat.

Providing high-quality, low-cost fortified complementary food is especially important in cities, where women sell their labour and buy food. Compared with rural dwellers, urban women have less-flexible schedules and tremendous time constraints, so they value convenience highly. They need high-quality, low-cost food to complement whatever breast milk they're feeding their babies after six months of age. Home-prepared complementary foods are good if they contain adequate nutrients. But home-cooked food often does

not have the nutrients a baby needs to grow healthy. Branded fortified infant cereals are widely available in major cities throughout the
developing world, but they are too expensive for most people, and
they are even less available and affordable in rural communities. The
result is devastating: nearly 40 percent of children under five in
these countries are stunted,[46] and nearly half of children under five
are anemic.[47]

Some outstanding initiatives are under way to combat micronutrient deficiency, including two founded in Canada, not far from our
home base in Toronto. The Micronutrient Initiative, launched by
Canada in 1992,[48] has developed a salt fortified with iodine and iron.
The iodine in this double-fortified salt aims to prevent thyroid goitre
in young children, which can lead to mental retardation, and the iron
serves to prevent iron-deficiency anemia.[49] This Ottawa-based initiative, with offices in New Delhi and Dakar and an annual budget of
around $30 million,[50] reports that it has reached nearly 500 million
people via health services. In addition to the fortified salt, it also distributes vitamin A in both capsule and syrup forms.[51]

Yet launching a fortification program is not as easy as one might
think. Consider the story of the Sprinkles® Global Health Initiative,
based in Toronto and led by our colleague Dr. Stanley Zlotkin, a
pediatrician and childhood nutrition expert.[52] A decade ago, Zlotkin
wondered why the majority of children in many developing countries were anemic, a condition that stunted their growth and delayed
their development. This wasn't a problem in North America, because
baby food manufactured in factories is fortified with iron and other
micronutrients. In poor countries, however, mothers made their
own baby food, and the staples they used—like rice or wheat—
didn't contain enough iron and other micronutrients. The answer,
Zlotkin thought, was to devise a system that would allow parents to

fortify food with iron at home. The challenge, he knew, was that many mothers in poor countries couldn't read, so the solution had to be simple—one dose in a single-use package. No measuring, no fuss. The iron also had to be in a form that didn't taste like metal.

So began years of testing at night in the kitchen at the Hospital for Sick Children in Toronto. "I gave Marcel, the head chef, a bottle of Scotch each month," Zlotkin told us with a laugh. "Then the hospital kicked us out of the kitchen, because of theoretical liability issues." Nevertheless, he managed to create a powder form of iron and vitamins A, C and D, together with zinc and folic acid, that he called Sprinkles®. There was just one problem: "I knew nothing about putting powder into packages." A colleague at the University of Toronto knew where he could go for help—the H. J. Heinz Company, headquartered in Pittsburgh, Pennsylvania. The ketchup and baby-food maker, which was looking for a project to fulfill its corporate social responsibility mandate, had extensive experience in putting condiments into little packages for the hospitality and airline industries.

Over the next decade, the Heinz Foundation would give Zlotkin $2 million to fund his research. Heinz not only designed little one-dose packages, it also got two of its subsidiaries to make and distribute Sprinkles® in India and Indonesia. The company also helped to write a guidebook on how to make Sprinkles® for groups that would manufacture it around the world. "It's not a hard sell," Zlotkin told us. "I go to some of the most remote parts of the world, rural communities that are literally at the end of the road. I ask mothers, what are your goals? They're always the same. They say: I want my child to be healthy, to have an education, to get a job. It doesn't matter where they are in the world. When I tell them their food may not be adequate, and if they use Sprinkles® it will probably improve their child's health, most people are willing to try it," he said. They can see the difference in their children right away. "With anemia, the ability of a young child to explore his environment is depressed. His development is depressed. In two weeks, the children change. They were lethargic before, and then they become more active. The women often say the same thing, with a twinkle in their eye: the children have become naughty. That's a normal child."

In 2007, the Sprinkles® Global Health Initiative created by Zlotkin and Heinz gave away its patent (except in the United States and Canada) on the invention so that anyone could make Sprinkles® and distribute it. Now it's being distributed in clinics or sold very inexpensively in stores in thirty countries in the developing world. Zlotkin reckons that about five million children are eating Sprinkles® right now. One current form of Sprinkles® is the equivalent of a multivitamin tablet, with fifteen micronutrients. Sprinkles® is working well in reducing illness, according to the Sprinkles® Global Health Initiative. In Bangladesh, one study showed that consuming sixty sachets of Sprinkles® with iron and other micronutrients over four months cut the prevalence of anemia by 65 percent. In the slums of Karachi, sachets containing zinc are significantly reducing the prevalence of diarrhea.

Zlotkin told us it had been a successful partnership with Heinz, but we could sense a hesitant note in his voice. The hesitation comes from his experience of the "general distrust of the private sector," he explained. "At times, this has tarnished the reputation of the program." UNICEF, for instance, works closely with breastfeeding advocates and has been a sharp critic of infant-formula makers like Nestlé. The agency was initially skeptical of the prospect of working with Heinz, even though the company makes baby food, not infant formula. Zlotkin overcame that reluctance; UNICEF is now an enthusiastic partner in encouraging children in the developing world to eat Sprinkles®. But leaders in some developing countries are harder to convince.

"I was just at a lecture in Bangladesh where I heard someone say, 'We don't want the private sector involved. They do a bad job. We have good food. Variety will solve the problem,'" said Zlotkin. "I said, 'Bangladesh has 80 percent anemia, 60 percent stunted physical growth, and it's been like this for many, many years. If you want your kids to be malnourished . . .'" He paused. "I have to control my un-Canadian feelings of being pissed off," he continued. "But I think these leaders are wrong. I've spent a lot of my time trying to justify the private sector's role in the production and distribution and actual making of the product. One can't have production without the private sector involved."

Zlotkin has been named to the Order of Canada for his pioneering work and is now the vice-president for medical and academic affairs at the Hospital for Sick Children—the same institution that once hounded him out of the kitchen. He has shown that working closely with a baby-food maker to fortify the food that poor people eat can do a world of good. Why aren't other baby-food makers teaming up with global health groups to address this pressing problem? Why can't they emulate the successful collaborations that are producing promising drugs, vaccines and technologies?

The answer lies in lack of trust.

To understand the depth of the mutual suspicion between health activists and multinational food companies, we need to return to the Nestlé saga, which continues to resonate, even to this day, every time the topic of fortifying complementary food for infants aged six months and older comes up. In the 1970s, any semblance of trust between Nestlé and the global health community was shattered by reports that the multinational food company was peddling infant formula to women in poor countries.[53] Different sources presented the facts in different ways, of course, but according to one source, Nestlé hired women with no special training and dressed them up as nurses to hand out free samples of its baby formula. "The samples lasted long enough for the mother's breast milk to dry up from lack of use," according to breastfeeding.com.[54] "Then mothers would be forced to purchase the formula but, being poor, they would often mix the formula with unsanitary water or 'stretch' the amount of formula by diluting it with more water than recommended. The result was that babies starved all over the Third World while Nestlé made huge profits from this predatory marketing strategy."

Reports like these led to a spectacular battle between Nestlé and global health advocates. To quote from a 1985 paper[55] on the international infant-formula controversy: "In 1974, a pamphlet entitled 'The Baby Killer,' was published by War on Want, a British non-profit organization. This report was extremely critical of Nestlé's operations in Africa. A Swiss group published a German-language version entitled 'Nestlé Kills Babies.' As a result, Nestlé sued for libel." Another report[56] states, "In June 1974, Nestlé sued the ADW [Arbeitsgruppe Dritte Welt] for libel on several counts, eventually narrowing them down to the title alone. The trial, which took place in Berne, Switzerland, was a classic mismatch: the ADW turned out to consist of 17 unknown activists who thoroughly enjoyed skewering the giant Nestlé in public for the two years' run of the trial. In July 1976 Nestlé won its case, but at a terrible cost in public relations." Nestlé won a Pyrrhic victory. The judge ruled that the company could not be held responsible for the infant deaths "in terms of criminal law"— but fined the defendants only 300 Swiss francs. Then, in a rebuke to Nestlé, the judge said the company needed "to reconsider its advertising policies to avoid being accused of immoral conduct."[57] The widespread publicity over the trial contributed to an international boycott of Nestlé that was launched in 1978.[58] Three years later, the WHO adopted the International Code of Marketing of Breast-milk Substitutes,[59] which prohibits the marketing of breast milk substitutes. The real target is infant formula, but the code catches complementary food in the net because, especially in its application to complementary feeding, it is broad in how it defines a breast milk substitute. It's hard to discern exactly whether the code applies to complementary feeding, and if so, how. There is no process for determining, for example, whether a particular complementary food, along with its marketing materials, would or would not constitute a violation of the code. And once a company is found to be in violation, there is no mechanism for a company to remedy or improve its processes and get off the list of violators.

The boycott against Nestlé ended in 1984, after the company agreed to implement the code. But four years later, the boycott resumed when the International Baby Food Action Network alleged that baby-milk companies were providing free samples to health facilities in the developing world.[60] As of 1997, a study[61] called "Cracking the Code," published by the Interagency Group on Breastfeeding Monitoring, reported that five companies—Gerber, Mead Johnson, Nestlé, Nutricia and Wyeth—were still handing out samples of infant formula to women in four developing countries. The companies disputed the charge, saying the study was "biased in design and execution,"[62] but it drove another wedge between industry and health activists.

This is unfortunate because the public sector needs help in many countries to tackle the massive problem of malnutrition. National authorities and international agencies such as the WHO and UNICEF are working hard to make fortified complementary food widely available, but in many places the products reach only a tiny fraction of those who could benefit because there are no effective delivery systems or outlets. What's the roadblock? Trust between the public and private sectors.

Multinational food companies could help the cause in plenty of ways. They could market their products with less-aggressive strategies that do not undermine breastfeeding and they could work hard to ensure that their local agents adhere to the WHO's code. They could work more seriously with local companies to build these companies' capacity to produce and market high-quality complementary food products. They could learn to talk more effectively to child health advocates and NGOs. In short, they could use their know-how in R&D, distribution and marketing to help develop and deliver complementary foods in the developing world. Anyone who has travelled in the developing world knows what we're talking about. You can get a Coke or a Pepsi in just about every spot on the planet, no matter how remote, but you often cannot get nutritious, inexpensive complementary food.

Multinationals, however, are reluctant to help. Of course they have their own commercial imperatives and global interests, but sometimes they are reluctant because they fear being attacked by NGOs. The NGOs contend that the companies are breaking the WHO code. The multinationals, as expected, deny these accusations. As a consequence, all of us concerned with the issues of malnutrition and under-nutrition are at an impasse. There is no trust, no problem-solving communication takes place, and children continue to suffer and die. The situation must not be allowed to continue.

How do we build the trust that will eventually save lives? No single method is going to work. What will be needed in the long run are many approaches aimed at building confidence between the public and the private sectors. Fortunately, there are a few rays of light, and some key players are now taking small steps. In 2008, GAIN, the Global Alliance for Improved Nutrition, received a grant of $38 million from the Bill & Melinda Gates Foundation to "work with private companies and public-private partnerships to introduce nutritious foods for infants and young children between 6 and 24 months of age."[63] Marc Van Ameringen, the executive director of GAIN, underlined[64] the important role the private sector can play: "The private sector can have a huge impact on the long-term health of infants and young children, who are at a critical stage in their mental and physical development, by producing healthy complementary foods specifically targeted at this vulnerable group."

As for us, we too have begun to work on this issue in a small way. In June 2008, Peter flew to Seattle for the Pacific Health Summit, an annual gathering of public and private sector leaders that is supported by the Gates Foundation. Peter's assignment was to chair a panel on the role of industry in the food crisis. The audience included NGOs, researchers and senior food executives. Amazingly, they had never before gathered in the same room to discuss how to address the devastating problem of childhood malnutrition. Clearly, they didn't trust one another. It was as if the NGOs spoke Latin and the company executives spoke Greek. Yet they all shared a common goal—to provide adequate nutrition to children in the developing world.

Everyone agreed that we needed to begin with small steps: a set of ethical values that could serve as a lingua franca, a common language for the various stakeholders. To build trust, we wanted to make explicit the principles that everyone, whether from the public or private spheres, could endorse. To that end, Peter made a public commitment to develop a code of ethical principles for infant and child nutrition. Our colleague Jerome Singh then studied codes of ethics of various organizations from both the public and private sectors to help develop these shared principles.[65] The principles we eventually adopted included vows of integrity, solidarity and fairness, as well as a promise to be responsible, and to comply with international codes and domestic laws. All sides have been asked to acknowledge the role of private enterprise "in scaling up production of low-cost, high-quality complementary foods and related products for infants and young children in developing countries in achieving our common goal."[66]

The president of the Gates Foundation's global health program, Dr. Tachi Yamada, committed to funding yet another confidence-building measure: a study that would use fair and transparent criteria to rank food companies on their concrete actions to improve access to nutrition in the developing world. Companies with a bad ranking would be exposed to public shaming. Companies with a good ranking, on the other hand, would be able to promote their positive actions. Such a ranking, called the Access to Medicines Index,[67] has been developed in the pharmaceutical sector. It has been well received, and is seen as having the potential to be effective in getting big pharmaceutical companies to make their products more accessible to people in poor countries.

A year later, at a smaller meeting, we came up with another creative way to build trust. The meeting had been called to focus on the WHO's marketing code—the maze of rules that were supposed to deter business from undermining breastfeeding. There were only twenty participants, and even before the meeting it was clear that stakeholders were deeply suspicious of one another. Nestlé and other violators of the code had not been invited; nor were the most strident finger-pointing breastfeeding advocates. But we did invite executives from food companies and public health activists who had never worked together before. Everyone wanted to do something to save millions of young lives, but the executives and the advocates were still speaking different languages. The exchange went something like this:

Food executive: If I want to market food, does the code say what I can and cannot do?

NGO technical expert: The code is not meant to limit complementary feeding.

Food executive: I understand, but let me repeat my question. For a particular food, for a particular marketing material, would I know for sure?

NGO technical expert: I guess it's ambiguous.

Food executive: Is there anywhere I can go for a determination on whether I would or would not violate the code?

NGO technical expert: No.

Peter: Wouldn't it be useful to have a place to go?

The solution to this dilemma came from an executive of an international clothing company who had faced a similar challenge and came up with a widely accepted way to work through it. This executive suggested that a panel could be formed at the global level for infant nutrition. The panel, which would include stakeholders from the breastfeeding activist community, international organizations and industry, would examine a particular complementary food and its marketing materials in advance and determine whether it was in compliance with the code. The food executive thought this was a good idea, especially if the global health activists respected this seal of approval. He said this would make it much more likely that multinational companies would be willing to invest in complementary feeding.

In fact, the current state of affairs in infant and young child nutrition is reminiscent of the anti-sweatshop debate of the mid-1990s, and in particular the criticism of Nike's use of child labour in the manufacture of soccer balls in Pakistan.[68] In that instance, too, there was mistrust and little collaboration between public and private sectors. Thanks to risk-taking by some stakeholders, including trade unions, advocacy organizations and company executives, there is today more collaboration between the public and private sectors to enforce child labour standards, including informal and formal complaint mechanisms to manage issues collaboratively when violations of working conditions occur.

A concrete example of a collaborative approach in the sweatshops debate that could serve as a model for infant and young child nutrition is the U.K.-based Ethical Trading Initiative, or ETI, a voluntary alliance of companies (more than sixty-five firms have joined, including The Body Shop, Marks and Spencer and Tesco), trade unions and NGOs concerned with labour standards. Their code of standards is based upon International Labour Organization conventions. When allegations of violations are raised, the complaint can be brought to the company or to the ETI itself. If the complaint is not remedied, stakeholders within the ETI can call for a commission to further investigate and arbitrate the complaint.

We strongly feel that these ideas—developing a set of ethical principles, ranking the food industry and encouraging multi-stakeholder initiatives—could go a long way towards building trust between multinationals and health activists who together have the potential to find solutions to the problems of nutrition in the developing world. We don't pretend these first steps are going to save the lives of several million children, but we do think they can create an environment that will enable key people to work closely together. Then these people can take the next steps themselves—to cement the trust and begin the work.

We do know that such trust *can* be built, even between former opponents. Indeed, we are seeing this for ourselves in our work developing a social audit for the Water Efficient Maize for Africa project, which aims to develop a drought-tolerant maize (corn) for Africa.

As we have seen, malnutrition and under-nutrition in the first two years of life is a massive problem. So too is the problem of hunger generally, among all ages, in the developing world. As of 2010, about 906 million people in developing countries—including one in three people in sub-Saharan Africa—are hungry.[69] Moreover, the number of hungry people in the world has recently increased because of the global economic and financial crises, and also because many more people in emerging economies are eating meat. Increased meat consumption in countries with growing economies means that land and water are used to grow cereals for cattle feed instead of food for humans, so the price of food goes up. Meanwhile, the local supply of food for humans is totally insufficient. While the yield of cereal crops in developed countries is over five tonnes or more per hectare of cultivated land, in sub-Saharan Africa it is just one tonne per hectare. This is the lowest in the world, and it hasn't improved since the 1960s. It's true that the Green Revolution of the mid-twentieth century helped improve yields in many parts of the world. In India, for example, farmers today are growing six times more wheat than they did in the early 1960s. Africa, however, did not implement the Green Revolution's farming methods, and that neglect meant that the revolution essentially bypassed Africa. Until now.

The Alliance for a Green Revolution in Africa, created in 2006 by the Rockefeller Foundation and the Bill & Melinda Gates Foundation, is trying to make up for lost time and opportunity by introducing agricultural improvements that will lift small-scale farmers and their families out of poverty and hunger. It is doing this through techniques such as improving the use of fertilizers, water and pest control, and improving breeds through conventional plant breeding. It's an African-led partnership, steered by former United Nations secretary general Kofi Annan.

This is a helpful, conservative approach that could significantly increase crop yields. But the conventional approach also has a ceiling on yields, a ceiling that can be broken only by using the power of modern life sciences. Science already offers a powerful tool to increase crop yields: the ability to create genetically modified crops that are better able to resist pests and withstand withering droughts. This could potentially help to save millions of people from hunger and poverty. But to do so, it seems that African countries need to work with Monsanto, the company that possesses not only the patents on many GM technologies but also the know-how to use them. Monsanto, the controversial seed company, has profited hugely from its genetic-modification technologies. It has introduced genes into corn and other crops so that they can withstand weed killers— herbicides that Monsanto itself sells. This technology enhances Monsanto's profits, but it also spares farmers the labour of weeding and allows them to use less herbicide than they did before. Monsanto has also developed plants that can fend off insects. To protect corn from the stem borer, which devastates crops, farmers traditionally used a naturally occurring pesticide, a bacterium called Bacillus thuringiensis, some other organic methods or, if they could afford it, costly and poisonous insecticide. Monsanto found a way of introducing the gene of B. thuringiensis into crops such as cotton, corn and potatoes to make them "insect resistant."

Now the U.S.-based agricultural giant has introduced technology to genetically modify seeds that can potentially help African farmers survive drought and disease. Monsanto is the leader in the race to make crops able to withstand a lack of rain. Its trials in the United States of a couple of drought-tolerant genes showed that they could improve yields of corn between 11 and 30 percent under drought conditions.[70] These drought-tolerant varieties of yellow corn were designed for lucrative markets in the United States, Brazil and Argentina—but not for Africa, where farmers of tropical white corn are too poor to be valuable seed-buying customers.

Then, in 2004, Monsanto offered to share its knowledge of drought-resistant genes for humanitarian purposes. The offer was not a pure act of charity, as Monsanto CEO Hugh Grant told Peter in the fall of 2009. They were chatting over dinner at the country retreat of the secretary general of the United Nations. Grant clearly recognized the technology's potential for small African farmers, but that wasn't the prime reason why Monsanto would make such an offer. He said Monsanto will be making significant profits with water-efficient technology in the United States and other rich world markets and that the company's ability to do so could be impaired if it withheld that same technology from the poor. This was an extremely insightful and honest statement. Grant knew he had to share the technology with the poor to maintain freedom to operate in rich markets.

But Monsanto faced a massive obstacle—widespread suspicion about its motives. The company was still struggling with the legacy of perceived arrogance that began during the leadership of Robert Shapiro, who had resigned in 2000. Under Shapiro's brash leadership, Monsanto had been attacked worldwide from many angles, for harassing farmers in North America, for spreading misleading information through advertising in France, for lowering levels of fertility in Austrian mice and for driving Indian farmers to suicide. Most of all, Monsanto was accused of an underhanded plan to dominate crops worldwide with a technology that produced plants with sterile seeds. Opponents dubbed this seed technology "Terminator" to evoke images of the Arnold Schwarzenegger movies of the same name.[71] This was a brilliant public relations move; the tagline from the original 1984 *Terminator* movie read: "In the Year of Darkness, 2029, the rulers of this planet devised the ultimate plan. They would reshape the Future by changing the Past. The plan required something that felt no pity. No pain. No fear. Something unstoppable. They created . . . 'THE TERMINATOR.'"[72]

This controversial seed technology would prevent seeds from spreading into the wild, but it would also require customers to buy seeds every growing season. Critics said this would bankrupt farmers who didn't have the money to buy seeds every year. In the end, the Terminator technology, which belonged to a company acquired by Monsanto, was never developed. In 1999, a year before Shapiro left the company, Monsanto promised not to develop sterile-seed technology in food crops.[73]

Since then, Monsanto has become a strong and highly profitable company, but its reputation has never fully recovered. Critics, including many activists in global health, do not trust Monsanto, even when it appears to be making a genuine effort to share its GM technology for humanitarian reasons. So after Monsanto offered to share its GM technology to develop drought-resistant crops in Africa, donors, government agencies and politicians in Africa continued to be suspicious of both the company and GM crops generally, even though the GM technology might significantly increase yields and food supply.

"It is discouraging," political scientist Robert Paarlberg wrote[74] in his 2008 book *Starved for Science*, "to see so few willing to help bring the latest in agricultural crop science—crops engineered with enhanced drought tolerance—to Africa's poor farmers." International donors were partly to blame, he wrote, yet it was African government leaders who refused to approve greenhouse and field trials of GM crops.[75] Paarlberg makes an important point. African politicians are worried that GM food will be hard to export to Europe, where suspicion of genetic modification is widespread. Many African politicians also don't trust a technology that's developed elsewhere. In particular, they're worried that U.S. corporations will exert too much control over their fields, seeds and crops.

We saw these concerns in action in Kenya in 2005, when that country was considering a biosafety law. During the hearings, local activists attacked the proposed law. The opposition hit its target. In the midst of deliberations over the law, the Kenyan agriculture secretary ordered the destruction of all corn crops that had been genetically modified to resist pests[76] (the crops were being tested in field trials). The agriculture secretary alleged, "There is an emerging tendency by our scientists of yielding to pressure from international collaborators pushing to secure approvals for their research projects faster, side-stepping procedures."[77] (Despite the opposition, Kenya finally did approve its biosafety law in 2009.)[78]

Here was a serious case of lack of trust in GM technology, and in Monsanto. And the mistrust was standing in the way of using Monsanto's water-efficient technology for the benefit of Africans. It was a roadblock, to be sure, but the Gates Foundation was determined to overcome it. In 2007, a Foundation official invited us to a meeting in Seattle. He knew we had studied Monsanto as part of a book we had written on ethics in the biotechnology industry. He wanted to know more about our perspective.

We told the official we thought the Monsanto brand was indeed a problem, but our observations had led us to suspect that the company might have changed, and might genuinely now want to use its knowledge for humanitarian purposes in Africa. In 2002, while researching for our book on ethics in the biotech industry (which we published in 2005),[79] we interviewed Monsanto's then CEO, Hendrik Verfaillie, and his senior team. They were honest and blunt about the company's failings. Verfaillie acknowledged that Monsanto had become the "personification of Evil" in Europe. Jerry Steiner, today Monsanto's executive vice-president of sustainability and corporate affairs, put it this way: "It was naive arrogance," he said. "We had a naive sense that our view would be unanimously accepted. We were arrogant by making decisions about what people would or should want."

Monsanto was clearly trying to do better. The "New Monsanto Pledge," instituted in November 2000, emphasized values such as integrity, dialogue and transparency. The company even published an annual "pledge report" to describe how it was doing in meeting its pledge commitments. Over time, the pledge report increasingly relied on data and perspectives from people outside the company. While some people continued to think this was a cynical PR exercise, by the late 2000s Monsanto was in the process of making three serious new commitments: to develop better seeds, to conserve resources and to improve farmers' lives.[80] Those promises, which were announced in 2008, made it clear Monsanto wanted to play a role in improving crops in Africa. This was further reflected by the following addition to their initial pledge: "We will share knowledge and technology to advance scientific understanding, to improve agriculture and the environment, to improve crops, and to help farmers in developing countries."[81] But pledges are one thing. Actually doing something is different. And the company's tarnished reputation had obviously not gone away.

In 2008, it was time to try again to unite Monsanto's expertise with Africa's need. The Gates Foundation and Monsanto announced that they had joined forces with African partners in a public-private partnership to introduce a form of genetically modified African corn that could withstand drought. The goal of this five-year project, named Water Efficient Maize for Africa, or WEMA, is to make drought-tolerant corn available royalty-free to small-scale farmers in sub-Saharan Africa.

WEMA is run by a non-profit organization based in Nairobi, the African Agricultural Technology Foundation. It is a classic public-private partnership. Monsanto is responsible, working with its African partners, for the genetic modification of corn to improve its water efficiency. Another non-profit partner, the International Maize and Wheat Improvement Center, is responsible for conventional approaches to plant breeding to select water-efficient corn seeds. Other partners are national agricultural research organizations, closely related to national agriculture ministries, in Kenya, Mozambique, South Africa, Tanzania and Uganda. The money to fund the enterprise, $47 million, comes from the American partners—the Bill & Melinda Gates Foundation and the Howard G. Buffett Foundation.[82]

As WEMA began, we knew this partnership faced a special challenge. In traditional agricultural regions, especially in Africa, indigenous food crops have deep cultural and religious significance. As a result, innovations affecting crops are viewed with distrust, which can increase the risk that agro-biotechnology initiatives will fail. Into this conservative world, WEMA was about to introduce Monsanto and genetic modification. We understood the risks all too well: if we didn't address public concerns appropriately, we would fail to deliver the benefit of science to the people who needed it. We advised the Gates Foundation that they needed to mitigate this risk, and they asked us to propose a new way to do this. The idea behind our proposal had emerged in our Seattle talk: we would conduct a "social audit" of the WEMA project.

Social auditing can be compared to financial auditing, in which an independent third party, often an accounting firm, examines the books of an organization to make sure it is reporting its financial situation accurately to its stakeholders. Social auditing, likewise, uses independent parties to assess an organization's performance against its stated social goals. The key difference is that while financial auditing uses widely accepted methods (so-called generally accepted accounting principles), the methods of social auditing are less developed and less commonly used. While social auditing has been widely used in areas such as child labour and environmental sustainability, to our knowledge it has not yet been used to assess global health or agriculture.

Our first act was to recruit a very capable and enthusiastic Nigerian runner, Obidimma Ezezika, who holds a PhD in microbiology from the University of Georgia and a master's in environmental management from Yale. His task was to develop a social auditing framework for WEMA based on the partners' own promises and agreements. Were they, in other words, doing what they had said they would do? WEMA's most important promise was to provide the water-efficient corn seed royalty-free to African farmers. They had also promised to listen to farmers and the public and to interact constructively with African agricultural researchers and national regulators.

We then identified a panel of about a hundred people in the five member countries to give feedback on how well WEMA was fulfilling its promises. The panel included scientists, farmers, government officials, parliamentarians, journalists, activists and others, including many people who knew about water-efficient corn and about the WEMA project but were outside of the project. The panel would be surveyed each year, and the findings shared with funders and posted publicly on the Internet.

When Obidimma first interviewed our panel in the fall of 2009, he found that views of the project varied widely. On ethical, social and cultural issues, the highest ratings came from regulatory personnel in the WEMA project, while the lowest ratings came from seed companies and farmers' associations. People's understanding of the WEMA project varied widely too. Strikingly, those with the least knowledge included academics, scientists, legal consultants and seed companies. The good news was that the panellists were keen to collaborate with WEMA to strengthen their understanding of the project. Academics and NGOs in particular want more information about the potential characteristics of drought-tolerant genes.

The findings give WEMA a chance to stay in touch with the sentiments on the ground, but they also suggest some potential actions the organization can take as it develops the drought-resistant seeds for Africa. Our first set of social audit findings led us to recommend, for instance, that WEMA make public the traits of the seeds that are being developed. WEMA should also share its thoughts on whether to add other traits to these seeds, such as pest and disease resistance. If this happened, would the seeds still be sold royalty-free? What's more, we suggested that WEMA do a better job of explaining its charitable purpose and describing how the intellectual property rights for these seeds will work.

We don't pretend that this independent auditing and reporting of ethical and social commitments in large-scale science projects in the developing world will make the ethical concerns go away, or will completely rebuild trust in Monsanto forever. But we do think our social auditing allows for the possibility that issues will be addressed honestly and transparently. Already, we see hopeful signs that the social audit is beginning to build a bridge of trust not only among the partners but also between WEMA and the community so that they can work together towards a solution. This will give the project a better chance to develop water-efficient corn, and to make sure it is understood, accepted and used by those who want it and need it.

As research continues on the drought-resistant corn, it's heartening to see that others have embraced the potential of some of Monsanto's genetically modified crops. On the outskirts of Johannesburg, Abdallah and Obidimma visited a small, well-kept farm, lush and green, with healthy stems of corn growing on the side of a small road. It was owned and run by a former boxer and freedom fighter, Motlatsi Musi, who had once worked as a farm manager for a white farmer. Now he was growing GM corn on his own small plot of land. He was well built, well dressed, articulate and very knowledgeable and passionate about agriculture and about GM technology in particular. After showing us his farm, Musi took us to a collection of small holdings nearby. (The other farmers had asked him to look after these properties for them.) He showed us one where a few rows of "Roundup Ready" GM corn had been sprayed with weed killer. The corn was healthy, and there were no weeds to be seen. In the next plot, he pointed to a few rows of corn that had not been genetically modified. It couldn't be sprayed with weed killer so, not surprisingly, it was clogged with weeds. In the next plot, conventional corn had been sprayed with weed killer by mistake; all of it had died.

We asked Musi if he was worried that anti-GM activists would come in the middle of the night to destroy his valuable GM crops. "No, I'm not worried," he said. "They would not dare to do that. They will need to consult with their undertakers before meddling with me!"

The same afternoon, we visited a woman in a small village on the outskirts of Johannesburg, where we could see Soweto in the distance. The houses here were mostly tin-and-mud shacks. Our host, a strong woman in a brown print dress, gold earrings and a snappy grey hat, was better dressed than her neighbours. She took us to a small plot of land she was farming behind the village. This was where people used to go to the toilet, she said. But now she's growing both conventional and GM corn on this bit of land. The conventional corn looked moth-eaten: it had been attacked by a worm known as the cornstalk borer. The GM corn a few feet away looked normal and healthy. It had been genetically modified to incorporate the *Bacillus thuringiensis* gene, which gave it the same protection against the stalk borer as the natural bacterium that some farmers have been using for ages to control pests. GM, this farmer told us, has made a huge difference in her life. Now she makes enough money from this little farm to support her family and help others. She has applied to the government to be given more land in a good location so that she can build a bigger farm.

Such examples are a start. These poor farmers are using modern biotechnology to improve their crops and create a better and healthier life for their families. If thousands more farmers benefited from the same GM technology that's used so broadly in North America, they too could improve their lives. But to make this happen on a large scale, public and private players need to drop their suspicions of each other and work together towards a goal we all share—ending hunger and malnutrition, and all the terrible health problems that accompany it. They need to build trust by participating in social audits and other confidence-building measures such as transparency rankings and ethics codes, and by involving all stakeholders along the way. We believe that trust is the critical barrier to making some key positive changes in global health, especially the development of GM crops to help combat hunger and improve nutrition for babies and young children. Public sector groups need to be convinced, with evidence, that companies like Monsanto are doing what they say they will do. Multinationals need to know they won't be skewered for trying to help and that they will be held to account if they misbehave.

It's a big challenge, especially after a history of conflict, but we think there are practical ways for multinationals to earn that trust, even from the most skeptical audience. Only by building trust can the public and private sectors work together to save millions from hunger, disease, death and disability.

CHAPTER SIX

We have shown how people around the world are thinking creatively about the scientific and ethical issues on the road from lab to village. Now we turn our attention to a third key barrier on this road: that of commercialization—turning the idea into a product that can be purchased and used by consumers.

In 2005, seeking ways to remove this barrier, we focused on Indian biotech companies in two major cities, Hyderabad and Bangalore. We started our studies of biotechnology companies with Sarah Frew. Sarah is an MIT-trained cancer biologist who began working with us when she realized that a traditional academic career does not provide professional rewards and incentives for young scientists who wish to work on global health–related projects. Within a few weeks of her joining the team in Toronto, we asked Sarah to begin planning her first trip to India to meet with local biotechnology companies.

Hyderabad, the capital of the Indian state of Andhra Pradesh, is famous for its cultural monuments and for the fabulously wealthy rulers, the Nizams, who governed the princely state until 1948. We went there to get a first-hand look at one of the most promising lines of attack against the diseases of the poor. On the outskirts of this bustling city, in a new industrial park that looks like it could belong in any American city, a cluster of biotechnology companies is on a mission to transform the health of a nation.

Peter was eager to see how these new biotechs were dealing with the profound health challenges of India and was soon climbing into a taxi, a ten-year-old compact without seatbelts, for the ride from downtown Hyderabad to the suburbs. As the driver gunned the engine and jockeyed for position with others honking their way through narrow streets under looming neon signs, Peter braced himself. The night before, on the way to dinner, his taxi driver had sideswiped another small taxi, and then jumped out for a furious argument in the middle of the bustling street. This time, Peter was lucky: the driver headed for the outskirts of the city without further incident. Soon he saw the sign: "You're Entering Genome Valley."

It is in places like this that we found one sustainable answer to the health problems of the developing world: domestic biotech companies that have drastically cut the price of existing vaccines and are beginning to invent new vaccines and drugs for the diseases of the poor in their own countries and beyond. You can see the future in leading-edge companies such as Shantha Biotechnics, which made its name by inventing a cheaper way to manufacture a vaccine for hepatitis B, a leading cause of chronic liver disease and liver cancer in the developing world. In the early 1990s, Shantha's innovation in vaccine manufacturing undercut the imported price of approximately twenty dollars per dose, and drove it down to less than one dollar per dose upon the launch of its own vaccine in 1997.[1] The company has since become a world-recognized name in vaccines.

Shantha is an example of the promising trend we're seeing in several emerging economies. They're not only creating affordable vaccines, drugs and diagnostics for the local population but they're making money while doing it. This is, in our view, a sustainable answer to the health problems of the poor, far more sustainable in the long term than charity. It is, we contend, more sustainable than some of the public-private partnerships we explored in the previous chapter, especially those driven by well-meaning Northern donors and multinational companies that have little engagement with companies in low- and middle-income countries.

Consider the prevailing business model to deal with diseases of the poor. It is a business model of multinational pharmaceutical companies driven by research and development. That business model is high cost, just like the movie business. Northern R&D firms lavish hundreds of millions of dollars on the discovery and development of a drug that may or may not succeed. Then the winners, the billion-dollar-a-year blockbusters, have to support the cost of the losers. As a result, the price of the successful drugs is very high. Northern R&D multinationals haven't changed their high-cost, high-price model to deal with diseases of the poor. Instead, they've sometimes given away their discoveries—either through straight charity or in partnership with philanthropists and the public sector. More recently they have begun to think about tiered pricing, whereby they charge poor countries less than the rich, although this does bring some theoretical difficulties such as re-importation of those products to rich countries. More fundamentally, this option does not address the need for emerging economies and developing countries to stand on their own feet and deal with their long-term needs through their own product development and commercialization.

The business model of biotechs like Shantha, on the other hand, is quite different. It's a model built on affordable innovation—more like Bollywood than Hollywood. These firms are keenly aware of the on-the-ground realities of local markets. They live in the communities that need these drugs, vaccines and diagnostics, so they are in a position to understand what people want and need. They can then respond with innovative products that people can afford. They can manufacture products at a fraction of the cost of imports because of low-cost labour and materials. And they have another key advantage: because these innovative companies are so close to their end users, they know how to market and distribute their product in a way that ensures that it makes a successful trip from lab to village. Their model, then, is low-cost production for low-income markets. For these firms, it's not an act of charity to develop drugs, vaccines and diagnostics for diseases of the poor. It's good business. Exhibit A: Shantha. When an 80 percent stake in the company was sold in July 2009, the company was valued at $784 million.[2]

A whole new generation of biotechs in countries such as China, India, Brazil and South Africa is now producing drugs, vaccines and diagnostics that cater to the diseases of the poor. When we studied seventy-eight companies in those four countries, we found some five hundred products for one hundred uses in the fight against disease.[3] While most of these products are small innovations—often more like adaptations than "new to the world" or fundamental innovations—the sheer number of products exceeds the number coming from public-private partnerships sponsored by huge U.S.

foundations and global pharmaceutical firms. These new biotech companies may not support Nobel Prize winners or be Silicon Valley start-ups with killer apps, but what they are doing, crucially, is developing, commercializing and marketing products that will significantly improve the health and the lives of millions of people in the poor world. They are developing products for local diseases such as Chagas disease in Brazil, rabies in India, hepatitis in China and HIV/AIDS in South Africa. They're also focusing on chronic non-communicable conditions such as cancer, cardiovascular disease and diabetes, which now kill more people than infectious diseases do in the developing world.

These local innovations may surprise many observers of the global health scene. When people think of pharmaceutical or biotech companies in a country like India, they usually think of masterful copycats. Until 2005, when India updated its patent laws to comply with the Trade-Related Aspects of Intellectual Property Rights agreement, it was perfectly legal there to essentially copy patented drugs.[4] An Indian company called Cipla, for example, became a hero of global health in the 1990s by copying drugs to treat HIV/AIDS, drugs that at the time cost individuals, on average, $10,000 a year. When Cipla offered the drugs for $350 per person a year to Médecins Sans Frontières, an independent medical and humanitarian aid agency, it forced the multinational pharmaceutical companies to slash their prices as well. The price cut likely saved millions of lives.[5] Today, low-cost manufacturing in India is a giant business. The Serum Institute of India in Pune, for instance, is one of the world's largest manufacturers of measles vaccine. It is not only India's biggest domestic vaccine supplier, it is also an exporter, with its products going to 138 countries, reaching about half of all children in the world.[6]

Now a second generation of biotechs is emerging in India and China. Both countries have signed on to a World Trade Organization deal to respect intellectual property, which means they can't simply copy other people's innovations that are still protected by patents. As a result, both countries are beginning to innovate, each in its own way. While many Indian companies are focusing on affordable vaccines, scientists in both countries—but more so in China—are leapfrogging to the forefront of genuine scientific innovation. They're now creating totally new drugs, vaccines and advanced molecular diagnostics. Somewhat controversially, some Chinese firms are among the first in the world to treat human patients with stem cell injections and gene therapy. So don't let anyone tell you that advances in biotech only happen in San Francisco and Boston. They also happen in Shanghai and Hyderabad.

This is the scientific example of Fareed Zakaria's "Post-American World." In his 2009 book of that name, the CNN host and *Time* Editor-at-Large argued[7] that the past five hundred years have been marked by three "tectonic power shifts." The first two were the rise of the Western world and the rise of the United States. Now, says Zakaria, "we are living through the third great power shift of the modern era," one that could be called "the rise of the rest."[8]

The rest, as Zakaria calls them, are challenging the notion that developing countries must wait for the developed world to make advances in science and technology that they later import at great human and financial cost. On the contrary, such countries understand that innovative science and technology is crucial to improving the health of their populations and their economies. Places like India, China, Brazil and South Africa are developing R&D infrastructures that will eventually break them free of their dependence on rich countries.

The story of Shantha in Hyderabad's Genome Valley is a prime example of this encouraging trend. When Peter visited Shantha in 2005, the company's headquarters was a stately white building standing alone in an oasis of trees and fountains. The surrounding scene was desolate—wide open fields and dusty roads. Only five years later, the area had been transformed into a well-kept industrial park with several brand-new buildings; it looked exactly like a campus you might find in a prosperous part of middle America.

This is where Shantha has built its business the classic way—by satisfying an unmet need. That need was an affordable vaccine against hepatitis B, a serious viral liver infection. It's easy to catch. Infants can catch it from breastfeeding or close contact with their mothers. Before there was a vaccine, many medical students and young surgeons would become infected by the time they finished their training. Most of those who are exposed to the virus become immune, but many develop a chronic form of the infection. About forty million Indians are thought to be chronically infected.[9] They can get inflammation of the liver, and then cirrhosis, a thickening of the tissues of the liver that resembles the damage from long-term abuse of alcohol. It can even cause cancer of the liver, which is almost always lethal. More than a hundred thousand people in India die every year from hepatitis B, but many more suffer from the other serious consequences, including massive bleeding from the esophagus and stomach, and chronic ill health.[10]

Shantha's founder, Dr. K. I. Varaprasad Reddy, was an electronics engineer with no medical training, so he didn't appreciate the scope of the hepatitis B disaster in India until 1992, when he attended a WHO conference in Geneva. Varaprasad was stunned, not just by the morbidity and mortality of hepatitis B in India but by the fact that a vaccine had been developed and marketed in the West over a decade earlier but still wasn't affordable for the poor in India, where it was desperately needed.

A hepatitis B vaccine had been developed at the Fox Chase Cancer Center, in Pennsylvania, in the late 1960s by Baruch Blumberg,[11] who would eventually share a Nobel Prize for the discovery.[12] In the mid-1980s, a genetically engineered vaccine to protect against hepatitis B came on the market.[13] Soon after, it became common practice in many wealthy countries for children to be vaccinated against the disease. But here was the hitch: the vaccines, made by monopoly patent holders GlaxoSmithKline and Merck, were expensive, at more than twenty dollars per dose. Most Indian families, who live on a dollar a day,[14] simply could not afford them.

What India needed, Varaprasad realized, was a homegrown hepatitis B vaccine at an affordable price. He set out to develop one, but it was very difficult to find the initial financing for a company to make the vaccine. He was rebuffed by every banker he visited, and one Western company even said that Indian scientists weren't sophisticated enough to understand recombinant technology to make the vaccine. That comment was like waving a red flag in front of a bull. Delivering an affordable hepatitis B vaccine became Varaprasad's life mission. He was so obsessed with it that he sold his father's property to raise $1.2 million. By 1995, he was on the verge of bankruptcy, but then he got lucky. The minister of state for foreign affairs of Oman injected $1.2 million in return for a 50 percent stake in the firm.

When Peter visited Genome Valley in 2005, he met Khalil Ahmed, a former employee of the foreign affairs minister who had helped persuade his boss to invest in Shantha. The minister had then sent Khalil to help run the company. Khalil, Shantha's executive director when we met him, has a sharp business mind, a refined manner and an exquisite taste for fine food, as Peter learned when he was invited to dinner at the Taj Hotel's private supper club. Over the meal, Ahmed was happy to relate the next chapter in Shantha's impressive business story. This story was later repeated to us in a beautiful seaside setting in Oman by the minister himself (with Abdallah helping to translate using his less-than-perfect Arabic, full of hilarious mistakes).

Two years after the Omani investment, Shantha produced India's first homegrown recombinant product, a hepatitis B vaccine called Shanvac-B. Although the vaccine itself had been invented by scientists elsewhere, Shantha came up with a cheaper way to grow the vaccine—in a yeast system. The key thing was that the price had to be right. "My gut feeling was that unless it is made for one dollar, nobody can afford this," Varaprasad said. Accordingly, Shanvac-B was initially priced at $1 per dose. It was an instant hit. Sales in 1997 were $1.6 million, far more than the projected $100,000. Consumption of the vaccine shot up from a few hundred thousand doses in the early 1990s to over thirty million doses in November 2008, thanks to the hard work of Shantha's 175-strong sales force, which marketed the vaccine mainly through doctors. It turned out that a high-volume, low-cost vaccine made for solid business. Profit margins were reportedly about 15 to 20 percent, and by 2009, revenues had reached around $90 million.

"It was perceived as impossible to develop a recombinant vaccine in this country," Varaprasad continued. "But still we did it because it was a noble purpose—to see that this vaccine goes to the public at a good price. So how did Shantha's scientists manage? I do not define it as God, but some unknown force."

It is important to note that it took more than thirty years from the discovery of the hepatitis B virus and the development of the first vaccine, and almost twenty years from the licensing of the first recombinant vaccine in the United States, for the vaccine to become affordable in the developing world, where most of the infected patients, and patients with hepatitis B–related liver cancer, are to be found. And this happened only because the vaccine was made in a developing country, with a low-cost business model, and through a process of innovation undertaken by an Indian company and driven by an entrepreneur who would not take no for an answer. It happened, but the odds against its happening that first time were tremendous.

After the successful launch of the hepatitis B vaccine, Shantha was determined to innovate further, initially reinvesting 25 percent (later 10 to 15 percent) of its profits into R&D every year—a higher figure than is typical of many other Indian companies in the same field. The goal was ambitious—to put a new product on the market every one or two years. "The criterion for development was simply to look at products that were relevant to India's and the other developing countries' needs," Shantha's former chief scientific officer, Ashok Khar, told us. To boost its homegrown efforts, Shantha developed partnerships with the U.S. National Institutes of Health, the Bill & Melinda Gates Foundation, Johns Hopkins University and PATH (the Seattle-based Program for Appropriate Technology in Health).

That strategy paid off. In 2005, Shantha launched India's first four-in-one combination vaccine against hepatitis B and routine childhood diseases that have been virtually eliminated by vaccination in most rich countries—diphtheria, tetanus and pertussis (whooping cough). Another Shantha vaccine targeted five strains of *Hemophilus influenzae* type b, a bacterium that causes severe pneumonia and meningitis in children under five, including two to three million cases of serious disease and about 386,000 deaths in 2000.[15] Shantha also launched a recombinant interferon alpha 2b product to treat chronic hepatitis. Selling at about $6.50, it is 75 percent cheaper than similar imported drugs.

In 2002, Shantha became the first Indian company to receive WHO prequalification, a regulatory requirement for selling a particular vaccine to UN agencies. It was by then supplying a significant proportion of UNICEF's recombinant hepatitis B vaccine, and by late 2009 it had landed another big contract from UNICEF, to supply $340 million worth of Shan5, a combination vaccine against diphtheria, pertussis, tetanus, Hemophilus influenzae type B and hepatitis B. Unfortunately, in early 2010 batches of Shan5 were recalled by the WHO because of quality concerns, thereby providing an excellent example of the ups and downs of pharmaceutical manufacturing. (The parent company is addressing the quality concerns and will likely reapply for qualification. Recently, a spokesperson was quoted[16] as saying, "Given the prequalification requirements, it could happen by 2013.")

Shantha's early success showed that scientists working in India were able to conduct advanced biotech R&D. It had a broader impact too. Since Shantha paved the way, other Indian companies have honed in on the business and are producing hepatitis B vaccines. The Indian biotech sector, almost non-existent in the early 1990s, was on track to generate about $3 to $5 billion in annual revenues in 2010.

In 2006, the Omani investor sold its stake to the French health care company Merieux Alliance. This new company acquired a 60 percent stake in Shantha partly because Christophe Merieux, the heir to the family fortune, was a physician who believed in vaccinating the world's children at an affordable cost. Just when the company was considering the Shantha purchase, Christophe died at thirty-nine of a heart attack. The company bought shares in Shantha as a tribute to him, and three years later sold its controlling stake to the French-based multinational Sanofi-aventis. This brings things full circle, in a whole new way: a developed-world company is buying a developing-world company that has become successful by taking the science of the developed world and making it work for the developing world. We are keenly watching to see how this type of dynamic will serve the interests of the poor in the developing world.

It is clear that Shantha's commitment to the developing world helped it to achieve its financial success. Its founder, K. I. Varaprasad, recognized that expensive multinational recombinant vaccines had reached only a tiny part of the market in India and the rest of the developing world, and therefore the full potential of the vaccine had not been realized. He saw that he could engage India's homegrown scientists, and use the lower cost of labour and process innovation, and a low-margin business strategy, to exploit this opportunity for his business and for the benefit of the poor. Varaprasad also led the way by finding international partners who could help the young company to grow. He invested in R&D from the very beginning, which helped Shantha become the first Indian firm to receive WHO prequalification for an Indian-made recombinant hepatitis B vaccine, opening the door to large international contracts.

What Shantha shows decisively is that a billion-dollar biotech company cannot only be built in the developing world; it can be built for the developing world. Shantha's affordable high-quality vaccines have already reached hundreds of millions of children globally. And Shantha is not alone. One of its biotech neighbours—and hepatitis vaccine competitors—in Genome Valley is Bharat Biotech International. The CEO, Krishna Ella, is a loquacious, pleasant man, with wide knowledge and interests and the gregarious manner of a local politician. He feels just as passionately as Varaprasad does about creating products for the poor: "If there is no innovation, there is no affordability," he said. "Affordability is linked to the innovation."

Bharat is teaming up with PATH to find a vaccine for rotavirus, one of the leading causes of childhood diarrhea and childhood death in the world. While vaccines for rotavirus developed by GlaxoSmithKline and Merck have recently been introduced, Bharat is working on a vaccine for a local strain of the disease. The so-called Bhan strain was isolated in India by the current director of the Indian government's Department of Biotechnology, M. K. Bhan. If successful, this rotavirus vaccine may be more affordable, and because it's targeted to the local strain, possibly more effective in India against this great killer of half a million children worldwide each year.

Bharat has also partnered with the Malaria Vaccine Initiative and the International Center for Genetic Engineering and Biotechnology to develop a candidate vaccine against *Plasmodium vivax*, the parasite responsible for most of the malaria cases in India. The money for this research comes from international donors. Bharat was the first biotech firm in India to receive grants—more than $11 million— from the Children's Vaccine Program (which is now part of PATH).[17] These funders are paying most of the bills for the company's efforts to pursue high-risk projects; Bharat only has to pay a quarter of the development costs. Under the terms of this public-private partnership, Bharat will sell the malaria vaccine for less than one dollar per dose. In return, it gets exclusive manufacturing and marketing rights. Most of the risk, in other words, is assumed by PATH, while Bharat benefits from access to the technological knowledge that will allow it to improve its capacity for innovative research.[18]

Bharat is an example of an Indian company that is inventing new ways to attack ancient infectious diseases in partnership with PPPs from Europe and North America. This kind of partnership allows everyone to benefit. There's real hope now for a lasting and sustainable solution to rotavirus and malaria, and under the terms of the deal successful products will be distributed at a price people in India and elsewhere in the developing world can afford. Furthermore, like other similar companies in India, Bharat is gaining valuable technological knowledge that will drive innovation in the future.

Genome Valley is not, of course, the only biotech cluster in this vast country. In order to get a more expansive perspective, we also visited Bangalore. It was here, three decades ago, that a twenty-five-year-old brewmaster launched a company that would become India's largest biotech, making everything from insulin to antibodies. In 2010, with more than four thousand employees, Biocon had revenues exceeding $500 million. Its CEO, Kiran Mazumdar-Shaw, India's "Biotech Queen" as *Forbes* called[19] her, is now one of the wealthiest women in India.

Mazumdar-Shaw greeted us in a conference room inside a gleaming, modern glass building whose entrance was graced with a modern sculpture representing the wheels of the industrial revolution. Mazumdar-Shaw is a warm and passionate CEO who clearly relishes the challenge of any obstacle that stands in her way.

The daughter of a masterbrewer, Mazumdar-Shaw followed her father into the business and even worked for the Guinness brewery empire in Ireland. She soon figured out that fermentation is basic biotechnology, and thought she could manufacture enzymes. The obstacles would have put off most people: biotech was almost unknown in India; the power supply was intermittent; bankers weren't interested. But Mazumdar-Shaw was not deterred, despite being a young female entrepreneur fuelled more by drive and vision than by actual business experience.

Mazumdar-Shaw plunged into the enzyme business, manufacturing pectin to start with and eventually statins to lower cholesterol. At the turn of the century, she made her next big move. The diabetes epidemic was taking off in India, especially in its fast-growing cities, and Mazumdar-Shaw could see the need for affordable insulin to control the disease. The market at the time was dominated by Denmark's Novo Nordisk and a handful of other large companies such as Eli Lilly. Their insulin, however, was priced too high for most people in India.

Mazumdar-Shaw was convinced that her enzymes business could help. Biocon's proprietary fermentation method could withstand India's intermittent power supply and allowed the company to make insulin at half the price it cost foreign rivals. What's more, Mazumdar-Shaw had the personal charisma to hire talented scientists who were just as excited as she was to create a new business model based on a new science.

When Biocon's cheaper insulin hit the market, multinational competitors cut their prices by 40 percent or more. But here was the wonderful thing: Biocon was still making a profit, even with its low price. It showed that Southern biotechs can create new and affordable drugs and make money at the same time.

Biocon's success spawned plenty of competitors and has helped to turn Bangalore into a bustling biotech hub, with companies like Strand Genomics, a bioinformatics company; GangaGen, a maker of phase-based antibiotics; and ReaMetrix, a maker of whole-blood-based diagnostics instrumentation. And now Biocon is innovating in a whole new way. It's entering the risky and expensive business of discovering new drugs and vaccines. It is now developing one of the world's first oral insulin products for diabetes, and is in the early stages of developing a low-cost antibody product to deal with certain cancers.

This is an exciting move, but it also presents Mazumdar-Shaw with a dilemma: how do you keep prices low if you've just spent large sums discovering and developing a new product? "Now, I know that I can never make this cancer antibody affordable to everybody in India, because I know there are lots of cancer patients who can't afford it," she told us. "I am saying to my guys, let's price it at a level where we can recoup our investment profitably. But don't make it so expensive where you plug your price along with the multinational price." She says she's still able to make a decent profit at a lower price because costs in India are lower. "I did all the clinical development in India. I am doing all the manufacturing in India and I am passing on that benefit to the patients."

Now Mazumdar-Shaw is on a campaign to persuade her fellow biotechs to innovate instead of just copying. "I truly believe that developing-world economies can pursue innovation far more cost-effectively than the developed world," she said at a 2009 meeting on biotechnology that we all attended at the UN secretary general's retreat outside New York City. Biotechs in developing countries, she says, are capable of delivering affordable outcomes that can address a number of unmet needs in health care, food, energy and environmental sustainability.

But how? Mazumdar-Shaw urged fellow biotechs to start simple and local, with technology they can handle, instead of deploying sophisticated technologies they may not have the skills to use. What's more, she said, they should partner with companies across the globe to expand research, development and marketing. Biocon, for example, has worked with a Cuban institute, thereby giving the embargoed country partnership opportunities that translate into capital and infrastructure for the benefit of both countries.[20] Biocon has also worked with U.S. companies to help them take their innovations to the market. In 2006, it bought a North Carolina company called Nobex for its intellectual property related to insulin delivery.[21] For all those who think it is only a case of companies in the United States buying companies in the developing world, think again.

It is entirely possible for biotechs in the South to start innovating. Yet many biotech companies in the developing world are reluctant to get into the risky business of discovering new drugs. Most of them "would rather slip into the cocoon of a comfort zone, imitating proven products and technologies," Mazumdar-Shaw said. "Consequently, most developing-world companies fail to think out of the box to exploit gaps in the marketplace. Most companies opt for low-risk services and generic versions of diagnostics, vaccines and therapeutics, involving process innovation focused on improving what has already been developed."

The conservatism of these companies is understandable. In India, it's harder to find the money to fund product innovation, as opposed to process innovation. To generate revenues, companies have little choice but to manufacture generics. Even Biocon did this with enzymes before hitting the jackpot with insulin and statins. There are so many barriers to innovation in India, it's not surprising that most of the country's hundreds of biotechs are either single-minded copycats or cautious process innovators.

Still, no matter where they are on the innovation spectrum, biotechs in India are successfully creating local solutions for local problems. Our research in India shows that the vaccines, drugs and diagnostics created by local biotechs are mostly targeting local diseases such as diabetes and diarrheal disease, while generating income, as in the case of Biocon, through the bulk manufacture of statins and other products for foreign markets. As we've seen, this is a far more sustainable model for the discovery and development of drugs for the poor than charities or public-private partnerships led by multinational pharmaceutical firms with headquarters and research and development departments on a distant continent. For most charities and PPPs, creating drugs for the poor is unidirectional: from the North to the South. For entrepreneurs in India and in China, it's a business.

It's hard to believe that only one generation ago, Shenzhen, China, was a fishing village. Today it's a booming city of approximately 8.9 million lit up twenty-four hours a day by screaming neon signs. Shenzhen, just north of Hong Kong, is one of the fastest growing cities in the world. It's also the home of Beike Biotechnology, a company that is using stem cells to treat everything from Alzheimer's disease and autism to cerebral palsy, diabetic neuropathy, arteriosclerosis and spinal cord injury. After just one decade in business, Beike is harvesting and selling stem cells, which have the potential to divide into specialized cells. Stem cells are a major area of research in China. Having published over twelve hundred scientific papers on stem cells in 2009 alone, China is the world's fifth-largest source of such papers. Stem cells are becoming a big business in the country, too. Beike, for instance, has eighteen labs, located in hospitals and government blood banks, and it claims to supply stem cells to more than two dozen hospitals. More than five thousand patients, including nine hundred foreigners, have been treated with stem cell injections in clinics in China.

China is one of only a few countries in the world that has allowed doctors to inject stem cells (other than in bone marrow transplants) into patients as a clinical treatment. In most other countries, the use of stem cells in this way is at the research stage, but Beike has, for better or worse, skipped over this step. Peter was eager to witness the clinical use of stem cells. On a sunny day in 2007, he stepped into a Shenzhen hospital, where a dozen patients from Australia,

Europe and the United States were sitting in a brightly lit modern ward waiting for injections, administered through a spinal tap, of stem cells derived from the umbilical cord. Peter was not allowed in here alone; a minder from the company escorted him from room to room to visit the patients.

The first patient was a woman from Hungary who had flown here to get stem cell treatment for her baby, who suffered from cerebral palsy as a result of birth trauma. Peter spoke to her in Hungarian, which he had learned from his parents; maybe, he thought, he'd get a more honest story if the minder wasn't listening in. Before the stem cell treatment, the mother explained, her baby could not lift his head on his own; feeding was always difficult because the baby would cry. Now she showed Peter how her baby, after the treatments, could lift his head from the bed, ever so slightly. She said the feeding had become easier because the baby was less rigid.

Peter moved to the next patient, a well-dressed fiftyish woman from Australia. She had a progressive degenerative disease of her spinal cord and cerebellum, the part of the brain that controls coordination of motor activity. "I haven't been this well in a long time," she told Peter.

How did she know the stem cells helped? "You can understand what I am saying to you now, right?" the woman replied. "Well, if you had spoken to me on the telephone a couple of months ago, you would not have been able to understand a word I was saying."

These are, of course, anecdotes. They don't prove anything. The patients might have felt better because they wanted to feel better, not because of the treatment. Or perhaps the patients felt better because neurological diseases can wax and wane. The best way to tell whether stem cell injections work is by doing a double-blind randomized trial, the gold standard in research. Patients are split at random into two groups. One group is given stem cells, the other a placebo. Neither the doctors nor the patients know who gets what. Then researchers study the difference between the two groups to see whether the treatment actually has an effect. Until a study like this is done, no one in the scientific world will believe that stem cell injections can improve the life of a baby who can barely lift his head, or of a middle-aged woman who has a disease that makes speaking a struggle.

In late 2010, a clinical trial reporting positive safety data for Beike's stem cell transplantation method was published[22] in the *Journal of Translational Medicine*. China ostensibly has a strong regulator, the Chinese State Food and Drug Administration, which models itself on the U.S. Food and Drug Administration and has strict rules about clinical trials for new drugs, vaccines and foods.[23] China is so serious about the credibility of its Food and Drug Administration that in 2007 it executed the head of the organization for taking bribes. In 2008, the U.S. FDA set up a bureau in China to help with regulation[24]—but not until its head was guaranteed diplomatic immunity.

Despite this apparently rigorous system of regulation, stem cells derived from the umbilical cord, cord blood and bone marrow slipped through a loophole. Injecting a person with stem cells was considered a clinical procedure in China, and the State Food and Drug Administration did not require trials before such treatments were approved. But without the evidence from clinical trials, how does anyone know whether or not they work?

Peter brought up the question of trials with the CEO of Beike, Dr. Sean Hu. Hu's critics, and there are many outside of China, say he is a scientific cowboy who's trying out an expensive treatment on patients without any real evidence that it works. Consider a 2009 *Economist* article titled "Wild East or Scientific Feast?" that highlighted our research on the regenerative medicine sector in China; the image accompanying the article was that of a Chinese snake oil salesman. Our research was led by Dominique McMahon, who is interested in the links between biotechnology and international development. McMahon has devoted her doctoral research to investigating developments in stem cell research in China, India and Brazil. Through the course of her work, she has conducted more than a hundred interviews with researchers, clinicians, ethicists and government officials. When we sat with Hu, he struck us as being no huckster, but a man who was passionate about helping his patients. He genuinely believes his remedy is so effective that it would be unethical for him to deprive some patients of the treatment in a clinical trial. "You didn't need clinical trials to show the Internet worked," he told Peter.

On the surface, this is an odd comment. Yet it is true that if a drug has a spectacular effect, you don't need to do the gold-standard comparison of a randomized clinical trial. When the first drugs to treat tuberculous meningitis were introduced, people survived, and that was sufficient evidence that the drugs worked; without them, the diseases were always fatal. Doctors didn't need to compare the survivors with a control group, because they knew the control group—those who didn't get the drug—would always die. But stem cell therapy is not an example of a spectacular, life-saving improvement.

There is another way to prove that stem cell therapy works, though, and Peter suggested this to Hu. The technique, developed at McMaster University in Canada, is called "n of 1" trials.[25] It works for patients with a chronic disease. The patient receives the therapy for a period, then a placebo. He can't tell whether he's on a drug or not. At the end of the trial, researchers check whether the patient was in better shape on the drug or off it. Such a trial might offer a reasonable compromise. But Hu just picked up his chopsticks to resume eating, and did not respond.

At any rate, Hu might soon be forced to test his therapy. In 2009, China announced[26] regulations that will close the loophole that exempts clinical procedures such as stem cell therapy from the rigours of clinical trials. Under the new rules, hospitals will have to either stop providing therapies that are unproven, or apply to the ministry for approval. It's far from clear, though, how effective the enforcement of these rules will be.[27]

Beike is just one example of the new generation of innovative biotechs blossoming in China with heavy support from government. This development may surprise many Westerners who think of China as one big copycat rather than an innovator. And it's still true that 90 percent of China's biopharmaceutical market is made up of generic copycats.[28] These are meeting a massive need for low-cost products for China's 1.3 billion inhabitants, who make up one-fifth of the world's population. For some companies, generic products represent a low-risk entry point into the industry.

But ever since China became compliant with the Trade-Related Aspects of Intellectual Property Rights agreement in July 2001,[29] compelling it to respect intellectual property, innovation in that country is taking off. With the return of the so-called sea turtles— Chinese scientists educated abroad—China today is developing a biotech industry that is like a baby dragon waking up. A Chinese company, for instance, has created the world's first cholera vaccine in pill form.[30] Even more startling, a twelve-year-old company called SiBiono GeneTech was the first in the world to develop and market a gene therapy product. SiBiono conducted clinical trials of Gendicine and won approval from Chinese regulators in October 2003.[31]

The founder of SiBiono, Dr. Zhaohui Peng, had worked in California in the mid-1990s. He is an intense and focused scientist, and he visited Toronto in 2008 for an international meeting we had organized called "Mobilizing the Private Sector for Global Health Development." Peng was eager to convince a packed lecture theatre that his gene therapy works. In just thirty minutes he ripped through one hundred slides of pathology pictures, figures, graphs and other data. When his time was up, he was still going strong, speaking quickly in English with a heavy Chinese accent.

The scientists at Gendicine, Peng explained, take a piece of human DNA, p53, which codes for a protein that suppresses tumours, and put it into a virus that acts as a delivery vehicle. When the virus is injected into a tumour, it transports the p53 gene into a cancer cell. Once the gene is incorporated in the DNA of the cancer cell, it starts making the protein that suppresses tumours, effectively issuing new instructions to stop the out-of-control replication of the cancer cells.

Peng's company developed the drug with about $20 million from the Chinese government, banks and a small investment from a private firm for its R&D facilities. In 2003, Gendicine won approval in China for the treatment of a type of head and neck cancer called nasopharyngeal carcinoma, which affects as many as 50 out of 100,000 people in Southern China (but is rare in people of European or Anglo-Saxon origin). It's a big killer in China, largely because the disease is often not diagnosed until it's advanced. By then, ten-year survival rates can be as low as 10 percent.[32]

Since Gendicine was approved for nasopharyngeal carcinoma, it has been injected into more than nine thousand patients, twelve hundred of them foreigners, who pay $60,000 for an eight-week treatment. The scientists at SiBiono have meticulously recorded the results. Five years after the drug was introduced, they released the results of a randomized, controlled clinical study that compared forty-two patients receiving both the drug and radiation with another group who received a placebo and radiation. According to an article[33] in the *Journal of Clinical Oncology*, Gendicine "significantly increased" the tumour control rate for people with nasopharyngeal carcinoma. After five years, although the number of patients in the trial was too small to rule out the play of chance, the survival rate of the Gendicine group was 7.5 percent higher than it was for the group that received radiation alone. This is admittedly not very impressive, but it follows the way cancer treatment improvements have occurred in the West, and shows that China has the capability to do scientifically sound research while innovating.

In his Toronto talk, Peng was obviously proud of the success of his novel gene therapy. One problem is that only a tiny fraction of the people who suffer from this form of cancer signed up for the treatment, probably because it cost roughly $15,000 for citizens of China, and all doctors administering the treatment had to be specifically trained to do so. Another problem Peng was facing was the fact that he couldn't export the drug because Western countries have stricter regulatory requirements.

In June of 2009, we visited the current chairman and CEO of SiBiono, Yiqing Wan, to see how things were going. Wan became chairman after a majority stake in the company was bought in 2007 by the Chinese company Benda Pharmaceutical.[34] We met Wan in SiBiono's clean, modern lab, which sits in the middle of a vast plain of neatly trimmed grass. Speaking through a translator, Wan said the company's scientists are working to improve the product's efficacy profile and expand Gendicine's scope to the treatment of other cancers. The company's scientists are working on this both in the lab and in clinical trials.

It's amazing to watch how quickly Chinese companies are moving to catch up with the biotech and pharmaceutical companies of the rich world. And it's heartening to see these firms attack some of the terrible killers in this immense country. Chinese health entrepreneurs, like the innovators at biotechs in India, are in a prime position to help their own citizens and make money by creating new and affordable drugs, vaccines and technologies for the diseases of the poor. They face plenty of hurdles. The financial system does not make it easy for international investors to sell companies and take their money out of China. There are ongoing doubts about the Chinese government's ability to control quality and clamp down on stealing ideas.

Yet there is no shortage of investors who are nonetheless eager to put their money into the rapidly evolving Chinese biotechs. Peter witnesses this from his vantage point as a member of the scientific board of a Chinese-based venture capital fund, BioVeda, which has plenty of American and international investors. One year the board was reviewing companies that distributed drugs; the next, they were looking at a company making monoclonal antibodies for eye therapy. The biotech dragon, in other words, is waking up. China doesn't yet have a huge and successful biotechnology company like U.S.-based Amgen, which rivals the size of some major multinational pharmaceutical companies, but we know that it soon might.

The Sinovac story is a prime example of the extraordinary speed of innovation in China. Sinovac was founded by Weidong Yin, a doctor who found his mission in the 1980s while treating infectious diseases at a time when more than 300,000 Chinese people a year were getting infected with hepatitis A.[35] Indeed, every year 1.4 million people worldwide are infected with hepatitis A through fecal-oral contact (the figure may be much higher since not everyone is tested).[36] Yin emerged from his experience in the field determined to create affordable vaccines for China. He made a vaccine for hepatitis A from an inactivated virus. Then, after obtaining permission to manufacture the vaccine, in 2001 he found a source of money—China Bioway Biotech Group and Peking University. They registered Sinovac as a business and created a Chinese entity capable of producing vaccines to combat infectious diseases in the local population.[37]

We discovered just how far Sinovac had progressed at the UN secretary general's retreat in the fall of 2009, at the height of the H1N1 pandemic. There was no H1N1 vaccine on the market in Canada and only small quantities of an inhaled vaccine in the United States; people in North America were understandably anxious. At the meeting, we met the head of the Chinese Centre for Disease Control and Prevention, Wang Yu. Sinovac, he told us, was one of the companies in China and India that had been given a sample of H1N1 virus by the WHO earlier that year to develop a vaccine. And yes, he told us, it had all gone smoothly: "Sinovac has made vaccine from the WHO seed stock. It has been approved by Chinese regulators. We found the vaccine was potent and did not need adjuvant. We are rolling it out to patients now." He sounded very casual about his country's success in producing an H1N1 vaccine, but we weren't entirely surprised—not after seeing the Chinese labs where focused, diligent grad students work around the clock. Sinovac became one of the first companies in the world to market a vaccine for H1N1 after its product was approved in China in September 2009. India's Zydus Cadila, based in Ahmedabad, had its H1N1 vaccine approved for marketing the following spring.

The process is clear. In both India and China, scientists began their work in the old-fashioned way, by copying. Starting with low-cost manufacturing, they improved the health of their local populations by cutting the price of vaccines, drugs and diagnostics. Then they made small innovations in manufacturing their products, which made those products more affordable, accessible and relevant to local and often global populations. Now, in some cases, they've developed entirely new products, a trend that we believe will grow in the near future.

Our studies show that biotechs in other developing countries are beginning this journey as well. In fact, the biotech industry is far more advanced in the developing world than most people realize. In our study[38] of seventy-eight firms in Brazil, China, India and South Africa, we found around five hundred vaccines, diagnostics and therapeutics that were being developed or were already on the market. These products were developed by local biotechs, not by local arms of multinational corporations. It was encouraging to see that over 120 of these treatments targeted neglected tropical diseases, as well as the Big Three—malaria, HIV and TB.

Diagnostics represent half of these products on the market and almost 40 percent of those in development. In Brazil, for instance, FK-Biotecnologia was developing point-of-care diagnostic testing platforms for Chagas disease, which occurs primarily in South America. In South Africa, Vision Biotech was developing rapid diagnostic test platforms for another local disease, African trypanosomiasis (sleeping sickness).[39] Point-of-care diagnostics mean that diagnoses can be made more quickly at the bedside, without the need for highly specialized technical staff and equipment found in centralized laboratories, which are often far away from the patients.

Vaccines, meanwhile, represent about 20 percent of the products on the market and over half of the products in development. The Indian company Shantha, for instance, launched an oral cholera vaccine, Shanchol, in early 2009. The vaccine was developed by the

International Vaccine Initiative, and Shantha has the licence to manufacture it.[40]

Drugs represent about 28 percent of the products from these companies on the market and about 7 percent of their products in development. Among those on the market are two formulations of amphotericin B for the treatment of human visceral leishmaniasis,[41] a parasite that attacks internal organs such as the liver and spleen and is fatal if left untreated, as well as a drug made by the Brazilian firm Hebron Farmacêutica—an extract from small-leaf mint that has anti-giardia and amoebicidal properties. Another Brazilian firm, Silvestre Labs, is selling a formulation of cerium nitrate and silver sulphadiazine, a broad-spectrum topical antimicrobial agent used to treat infection in skin wounds including those from leprosy. Several companies are selling anti-rabies immunoglobulin (antibodies) for protection following exposure to the rabies virus.[42]

Regrettably, despite early successes and the clear demonstration of the real and potential impact of these firms, these Southern biotechs are still not well known in the global health community. The PPPs in the West that are trying to invent new drugs, vaccines and diagnostics for diseases of the poor have thus far mostly overlooked the potential contributions of biotechs in the developing world. We think they should tap into the expertise of these Southern biotechs. These developing-world companies are well positioned for the battle against diseases of the poor. Close to consumers, they understand the local market and health delivery infrastructure. They have the potential to invent and develop drugs at far lower cost than Northern biotechs can. They're deeply invested in overcoming these challenges. As a result, a number of these firms have become "disruptive innovators." They're developing simpler and cheaper products and services for poor patients who have been previously excluded from conventional markets.

A few developed-world PPPs are seizing the opportunity. The Geneva-based Drugs for Neglected Diseases initiative, for instance, has teamed up with Brazil's Fiocruz and Bharat in India to develop malaria and rotavirus vaccines, with funds from the Seattle-based PATH. PATH has also teamed up with the WHO and the Serum Institute in India to manufacture a meningitis vaccine, which was launched in 2010. Yet these examples are exceptions rather than the rule. Right now, the Southern biotechs remain a largely untapped resource.

Clearly, it's much easier and cheaper to copy a drug, or to make a minor change in how it is manufactured, than it is to invent a drug from scratch. And to be sure, many of the biotechs in countries like India will continue to focus on small changes to processes that will considerably cut the cost of drugs and allow them to be manufactured in a profitable way. This is valuable, but it won't do anything for diseases for which doctors currently have no effective prevention, diagnosis or treatment. Vanquishing such diseases requires radical innovation and novel solutions for unmet needs. Some diseases will require purely scientific or technological innovations; others will require business or social innovations at the commercialization and delivery end of the road.

Scientists and entrepreneurs in the developing world, however, cannot join in the search for solutions without money and know-how—two things in short supply in most developing countries. We need to support indigenous biopharmaceutical firms in emerging countries on both counts. In the know-how department, GlaxoSmithKline is helping enormously by making public more than thirteen thousand promising compounds against malaria and contributing to a patent pool aimed at increasing access to proprietary technologies for global health.[43] Other companies are sharing their knowledge as well; it's a tremendous boon to research in the developing world.

We suggest three additional ways to help Southern health entrepreneurs deal with financial and other business challenges. To begin, countries like India and China could follow the example of the American "orphan drug" legislation, which provides a host of incentives for products targeted at markets with few people. These incentives could include speedier regulatory approval, longer patent protection, priority-review vouchers and even government procurement of the products. This could be a promising model for the diseases of the poor.

We've also proposed a way to help Southern biotechs export their products to distant markets. We call this approach the Global Health Accelerator. It would help health entrepreneurs in emerging economies, particularly the small and medium-sized enterprises, assess regulations in foreign countries, find commercial partners and distribution channels, and gain access to financing. In addition, it would include a global health prize to recognize excellent examples of Southern innovation against diseases of the poor.

Finally, entrepreneurs need venture capital, which has played such a critical role in the development of the U.S. biotechnology industry. Venture capital providers bring with them a range of strengths, such as skills in identifying promising opportunities, a willingness to invest in opportunities too risky for the banks, managerial support for the development of new firms, and networks to connect new firms with markets and mentors. Is the venture capital approach relevant to the developing world? Is it relevant to global health?

We think so. Venture capital investors could identify promising opportunities in the developing world arising from basic R&D and transform those opportunities into viable products and services. In this way, venture capital would supplement public funds, encouraging innovative local companies to solve local health problems. This model is already beginning to be realized.

The first venture capital firm, or VC, to invest in health R&D in China was co-founded by two "sea turtles," Harvard-educated Dr. Zhi Yang and Damien Lim, who trained at the University of Houston.[44] Between them, Yang and Lim have almost forty years of biotechnology and private equity experience in the United States and Asia, and they have invested in seven life sciences start-ups. With money from the International Finance Corporation (the IFC is the private investment arm of the World Bank), and investors in Singapore and Switzerland, they set up a $32-million fund, BioVeda China, and invested in a dozen companies in biotechnology, biologics, logistics and medical devices in China. Napo Pharmaceuticals, which is developing crofelemer, an innovative drug to treat chronic diarrhea; NOD Pharmaceuticals, which is developing nanoparticle oral formulations of insulin; and CITIC Pharmaceuticals, a medical-devices distributor, are examples of the companies that BioVeda has in its portfolio. Some of BioVeda China's companies from its first round of financing have since gone public.[45]

The success of the first fund led BioVeda to launch a second fund, with $100 million. They've found U.S. banks willing to invest, and have set up a scientific advisory board to deal with scientific and ethical issues (Peter is a member). It has also obtained financial and political support from the local Shanghai government.

Although the fund is for-profit, BioVeda China serves as a pioneering model of how international VCs and larger institutional investors can invest in a gateway to R&D in emerging markets. First, it shows how a for-profit fund in an emerging economy with a strong research base can create new health technologies. What's more, focusing on a diversified portfolio of firms with existing revenue streams, as BioVeda has done, can pave the way for larger, more high-risk innovation-focused funds.

There are several, but not many, other VC funds that invest in health technologies in the developing world, and each has a slightly different model. This makes sense, since the environment in each country is specific, as are the conditions in which innovative companies need to grow. In India, the first VC fund focused on biotech is the Andhra Pradesh Industrial Development Corporation–Venture East Biotechnology Venture Fund (APIDC-VE). It was founded in 2003 with funding from the Andhra Pradesh government and the Indian government's Technology Development Board, as well as key foreign investors such as the IFC, the Norwegian government and CITCO, a global financial services company. Its aim was not just to commercialize Indian biotechnology investments in a for-profit way but also to have a ripple effect of broader domestic benefits. One of the first investments was in Neurosynaptic, a firm that makes tele-medicine diagnostic equipment for rural areas. With a rapidly expanding network of over 150 centres, Neurosynaptic could potentially grow into one of the largest such networks in the world. It was named one of the World Economic Forum's Technology Pioneers of 2008.

The first life sciences VC in sub-Saharan Africa is led by a cell biologist, Heather Sherwin. As CEO of Bioventures, a $12-million fund, Sherwin has invested in eight companies in South Africa that have an R&D focus on medical devices, drugs and natural products, companies such as DISA Vascular, a maker of cardiovascular stents for coronary and peripheral vascular disease. While DISA has been a success, results overall have been mixed, partly because the fund is too small. Under pressure from investors to chase higher returns, Sherwin had to sell promising research to international investors rather than investing in R&D for locally endemic diseases. One investee, for instance, had a promising HIV/AIDS drug discovery platform in addition to a pain medication program. Yet the fund couldn't afford to keep putting money into R&D. "We canned the AIDS program in favour of pain and oncology programs whose intellectual property could be sold to international investors," said Sherwin. "If you follow your pure VC mandate, you cannot invest in TB or HIV companies. You're not going to get your returns."

Still, Bioventures' strategies could be a model for funds that aim to have a social impact. Bioventures provided companies with well-connected board members, strategic services and sympathetic feedback in difficult times. It sought out exciting R&D, rather than waiting for business plans to come to it. But in general, these funds often struggle to balance the potential for profits against the benefit for society. They need the metrics to measure the social outcomes against the financial ones. Metrics help answer the key question: what are funds achieving for the money put into them? There are metrics out there, such as Acumen Fund's Best Available Charitable Option. It's been used to evaluate investments in an African mosquito bed-net factory (which we will look at in the next chapter). It compares the results of investing directly in a bed-net factory with simply buying and distributing bed nets. But metrics are not enough. These VC funds need enough long-term (patient) money to invest effectively.

To overcome these challenges, our colleague Hassan Masum, alongside others, has proposed the creation of a "fund of funds." Hassan is an author, co-founder of an educational start-up and contributor at WorldChanging.com. He has collaborated on numerous scientific papers with us, and worked with government agencies, research labs, start-ups and non-profits to tackle complex socio-technical challenges. If a fund were capitalized at several hundred million dollars, it could provide matching money for the creation of several regional venture funds of $20 to $50 million each. Each regional fund would be run autonomously by professional venture capitalists with expertise in the health sector. The large size of such a "Southern Health Venture Fund" would allow it to make additional capital injections as needed, addressing capital shortages that have plagued funds like Bioventures. This has been tried before, notably by the IFC, but only on a case-by-case basis, not as an overall plan. The success of some of those cases, like BioVeda in China, APIDC-VE in India and Bioventures in South Africa, suggests to us that an overall approach could work well.

A fund of funds might acquire money from development banks like the IFC, and from philanthropies and governments. Since the timeline to take a novel therapeutic to market is typically a dozen years, longer than the average fund life of a VC firm, the fund's initial pool of capital would likely come from non-profit organizations such as local governments or foundations. Investments in each regional fund might be syndicated with experienced and well-connected international venture capitalists. One example is the recently created $150-million Malaysian Life Sciences Capital Fund, which teams up the Malaysian government with the San Francisco–based venture capital firm Burrill & Company.

We think a fund of funds approach could mobilize enough money to be effective. The network of regional funds could benefit from a peer review process, both between regional funds and with syndicating VCs. In the long term, these health-oriented Southern VCs can contribute to a more sustainable approach to global health innovation. They can stimulate Southern solutions for Southern health problems and create a virtuous cycle of economic development and high-quality jobs in the South.

By bolstering health entrepreneurs in developing countries like India and China, we could take a significant step towards solving a big chunk of the global health puzzle. Biotechs have already shown they have the expertise and the willpower to produce life-saving products at a fraction of the rich world's costs. Now that they're starting to invent new drugs, vaccines and diagnostics, they will be key players in the fight against neglected diseases. And the time to act is now. Biotechs like Shantha and other Southern companies owe their success to their ability to make drugs affordable. Now that Shantha is in foreign hands, will it still care about diseases of the poor? Or will it veer towards more lucrative markets? The same question could be asked of other biopharmaceutical companies that are faced with increasing costs associated with the development of innovative products. Will they try to recoup their investments by targeting diseases of affluence? In other words, which will they choose: global health or global wealth?

This very question was posed by our colleague Rahim Rezaie in a commentary in the September 2010 issue of the leading journal *Nature Biotechnology*. Rezaie left his job at a hospital laboratory in Toronto, where he had worked on diagnosing a host of genetic disorders, to pursue a master's degree in biotechnology. He thought this would be a way to become involved in a company that tries to develop health products for the poor. When the opportunity arose to study many companies involved in this mission, "I just could not resist the fusion of passion, curiosity and interest that this project presented—not to mention tourism," he says sheepishly. Since 2006, Rahim has visited about fifty companies in more than a dozen cities across China, India and Brazil, and he has interviewed scores of other people in research institutes, government agencies, industry associations and venture capital funds.

Rahim's work, as well as that of other members in our team, suggests that local firms in these countries present a tremendous and growing resource for global health. But leaving them alone to deal with the push and pull of market forces puts their long-term potential in this regard at great risk. With a bit of targeted support, we think emerging-market firms can affect both global health and global wealth. Proper support mechanisms will require concerted action on the part of the global health community, including governments in emerging economies and international donors.

Northern players need to engage emerging biotechs in the search for health solutions. They also need to support Southern entrepreneurs, with some of the measures we've proposed. Doing so will lead to healthier and wealthier countries in emerging economies and other developing nations. If we do this right, we will achieve something truly significant and life-changing for millions: sustainable Southern businesses that will solve many of the challenges facing their own and other developing countries.

CHAPTER SEVEN

How do we keep poor countries poor? This is easy to accomplish: ensure no domestic ideas are turned into products, sold locally or exported. Give the countries handouts. Don't let them develop their own solutions to their problems. And if someone in that country gets too much scientific training, if he or she is too ambitious or capable, make sure that person emigrates to a rich country. Sound familiar? This is precisely what is happening today in much of the developing world, and especially in Africa.

There is no greater gap between perception and reality than the difference between the common outsider's view of Africa and the view from inside. The word *Africa* itself probably conjures up in many readers' minds malnourished, pot-bellied, barefoot, starving children with doleful eyes, children such as those featured in late-night TV commercials soliciting contributions for charities. Don't get us wrong: these children exist, and the charities are laudable, especially in the case of acute humanitarian needs. But these images are no more typical of the people in the fifty-three diverse countries in Africa than images of cowboys are typical of Americans.

Our experience of Africa—experience that includes visits to, and studies in, Ghana, Tanzania, Rwanda, Uganda, Nigeria, Kenya and South Africa—has given us a completely different view. We see tremendous ingenuity, ambition and entrepreneurship. We see it in thousands of small shops and restaurants, often by the side of the road, that are teeming with customers. We see great wealth and poverty, luxury and privation. The people we meet are warm and beautiful. Their children are curious and mischievous. The landscape, and the wildlife, is majestic. Africa, we contend, is a misunderstood continent.

Few people realize that in the last few years many sub-Saharan African countries have had GDP growth rates that approached those of China and India. And even during the global recession, the continent's GDP grew by 2 percent, which was more than anywhere else in the world apart from the Asian powerhouses. Here's another surprising statistic reported recently by the *Guardian* newspaper:[1] on a per capita basis, Africans are already richer than Indians, and a dozen African states have higher gross national income per capita than China. Surprisingly, most of the growth is fuelled not by the sale of diamonds and oil but by the purchase of goods and services by Africans themselves. A middle class of accountants, teachers and vendors is rapidly emerging in Africa, just as it has in China and India. This African middle class may include as many as 300 million people, nearly one-third of the total population, according to Vijay Majahan, a development expert and author of the 2009 book *Africa Rising.*

Africa, in other words, is moving into a position where it can begin to solve its own health problems rather than simply being the passive recipient of money and medicines to save lives. This, then, is a story of hope, with a very different storyline than the one you usually read about this continent. We believe Africa can solve its problems through its own innovations. This is already happening in Tanzania, where a company called A to Z Textiles has become Africa's biggest manufacturer of bed nets impregnated with long-lasting insecticide. These nets are saving huge numbers of children from death by malaria that is spread by mosquitoes. Yet A to Z does more than make nets that ward off or kill mosquitoes; it employs more than six thousand Africans, lifting them and their families out of poverty and protecting them from the diseases associated with poverty. In A to Z's sprawling manufacturing plant in the beautiful mountainside city of Arusha, we see hope for the future of Africa.

But A to Z's success is the exception rather than the rule. The problem does not lie in the ability of Africans to innovate; in the continent's most economically and politically stable countries, the seeds of scientific innovation have already been planted. These seeds, however, need to be nurtured so that they blossom into practical innovations that improve people's health through earlier detection of disease, new drugs to remedy problems and even vaccines to prevent illness. Yet in most places in Africa, scientific ideas, even great ones, still go nowhere. They're never commercialized. People in the business of improving health in Africa don't even know about helpful innovations in their own countries. This is a terrible waste of talent and of ideas that could provide lasting solutions to some of the big health problems in Africa.

Here's a prime example. Professor Kwabena Bosompem works at the Noguchi Memorial Institute for Medical Research in Accra, Ghana.

The institute, named after Japanese medical researcher Hideyo Noguchi, was founded in 1979, and construction of the facilities was funded by a grant from the Japanese government. Noguchi died in 1928 of yellow fever in Accra, and to mark the eightieth anniversary of his death the Japanese government established the Hideyo Noguchi Africa Prize, an international medical research and services award.

In his research lab at the Noguchi Institute, Bosompem set out to create a simple test for schistosomiasis, a waterborne parasitic disease that infects the bladder, liver and other parts of the body. Without treatment, the infection can scar the bladder badly and even lead to cancer. The gastrointestinal form of the infection can damage the liver and be lethal. According to Bosompem, the urinary type of schistosomiasis is so common in Ghana that children only fifty kilometres from the university play a game to see who has the bloodiest urine as a result of the infection.

The simple dipstick test that Bosompem developed is an inexpensive way to diagnose the infection early so that patients can be treated with an effective, cheap and easily available drug such as praziquantel. Yet Bosompem's handy device is sitting in his lab, unused in Ghana.

You might think Ghana would be able to commercialize such an innovation. Its government has set up a $19-million venture capital fund, led by a Ghanaian who formerly worked with the investment bank Morgan Stanley in New York. What's more, Ghana's recent— and charismatic—health minister, Major Courage Quashigah, strongly believed in commercializing health products. (Quashigah died in early 2010.) The country also has a strong regulatory agency. Yet despite all these advantages, Bosompem could not get his schistosomiasis dipstick test out of the laboratory and into the villages where the children needed it. A system to develop and commercialize innovations simply does not exist. When we heard this story, we began to appreciate what Professor Bosompem and African scientists like him were up against. We began to refer to technologies stuck in laboratories as "stagnant technologies."

Regrettably, there are plenty of examples. We visited many research institutions and universities in Africa between 2007 and 2010, in work spearheaded by one of our colleagues, Ken Simiyu, trying to identify technologies that have similarly stagnated in laboratories. A veterinary doctor from Kenya, Ken also trained in business management and worked both as a research scientist for a research institution and a marketing agent for a major pharmaceutical company in Kenya. In a way, he had shared the experience of many African scientists while working as a research scientist, but his desire was to capture the experience of fellow African scientists in their own words.

In Kampala, we visited a traditional plant-medicine centre. The scientists we met were very formal at the beginning of the meeting, until they started to talk about their products. They had plants for everything—from sickle cell disease to sex. (The director, with a twinkle in her eye, assured us that a particular plant had improved her sex life.) Yet with an overall annual budget of less than $100,000, the centre couldn't afford to do much research, apart from some basic characterization of some of the plants and their active ingredients. No Western pharmaceutical or nutrition company has ever sat down to discuss a partnership with them. We do not know how many of their plants work, but we can't help but think there must be hidden gems among the ones being studied.

In Uganda we found another stagnant technology. The WHO's global program of childhood immunizations has created an unexpected problem: what to do with medical waste, particularly used plastic syringes. This has turned into a thorny issue in the villages because rural hospitals cannot afford conventional incinerators, let alone the fuel to run them. A professor in the Faculty of Technology at Makerere University in Kampala, Dr. Moses Musaazi, has invented an ingenious solution—an incinerator fuelled by the medical waste itself. The drum-like incinerator is being used at Mulago Hospital, the main university teaching hospital in Kampala, and two other hospitals around Kampala, but nowhere else in the country. Although certified by the World Health Organization, the technology has not been taken to the next levels—business planning, manufacturing, marketing—because the innovation system is not working.

Here's another example, from Tanzania. Artemisinin, currently the mainstay of malaria treatment worldwide, is extracted from the *Artemisia annua* plant. Used for thousands of years in China to treat fevers, it grows well in east Africa, but so far no one there has taken the necessary steps to manufacture the drug. What do they do? They extract a crude form and ship it to Novartis, a multinational pharmaceutical company in Switzerland. Novartis manufactures the drug and sells it back to the countries where the plant grows.

Artemisinin could be manufactured in Africa. A scientist in Tanzania contends he has discovered a novel and more efficient way to purify the key ingredient, but he can't commercialize his method—again, because the necessary elements of a working innovation system are not in place. This scientist is afraid to talk openly about his invention because obtaining a patent is so difficult, and without a patent, he worries that someone might steal his idea. This is a tragic waste.

Another Tanzanian, Wen Kilama, is a tall, distinguished, Shakespeare-quoting scientist who once directed the National Institute of Medical Research. He has spent a quarter century studying malaria. He has studied the mosquitoes that transmit malaria, the insecticides that repel them and the size of the pores a bed net requires to effectively protect children sleeping underneath them. And what is the result of Kilama's distinguished research? Some of it undoubtedly has informed health policy. And he has written articles that have been published in scientific journals. In the publish-or-perish academic culture that exists in Africa as elsewhere, this is the goal of scientific endeavour, tragically mimicking the academic culture of Africa's colonial powers. Indeed, we have encountered many professors in African universities with a monomaniacal focus on publishing their work in a prestigious journal. But that's where the ideas stop.

In Tanzania and in many other African countries, there is an unacceptable gulf between the scientists in universities and business people. A classic example of this divide can be seen in Tanzania, where Wen Kilama's impressive research exists alongside the Tanzanian business A to Z, which makes bed nets that are saving hundreds of thousands of children's lives—but without having used Kilama's research.

To see this made-in-Africa innovation, we drove to the outskirts of Arusha, where A to Z has an immense factory. Binesh Haria, the company's chief operating officer, and his cousin Anuj Shah, the CEO and son of the founder, met us in the impressive modern lobby. Haria hadn't had a day off in six months. He looked like he was midway through a marathon, while Shah exuded the cool air of a diplomat, covering up months of equally hard work. Haria led us into the factory—the size of a football field—and told us the incredible story of how the company got into the business of life-saving bed nets.

A to Z used to make T-shirts and other clothing, as well as conventional bed nets. But the old bed nets had no insecticide in the fibres. You had to douse them with insecticide, which came out in the wash, so the nets had to be doused again every time they were washed. The companies that made the insecticide preferred to sell it by the barrel, which meant that it was stored at a central depot. This was inconvenient for customers, who lacked both the money and the time to take the nets to a depot for redousing. The nets, then, were not very effective, and children continued to get malaria—and a great many of them died.

In 1998, Professor Don de Savigny of the Swiss Tropical Institute, who was then working with Canada's International Development Research Centre, flew to Arusha to find a Tanzanian company that would manufacture, distribute and promote a new home treatment kit for dousing bed nets with insecticide. He and Dr. Hassan Mshinda, the former director of Tanzania's Ifakara Health Research and Development Institute, initially approached A to Z's competitor, Sun Flag, but it wasn't interested. Then Mshinda suggested they try to sell the idea to A to Z Textiles, which was just across the road. They made their pitch to Haria and Shah. The home treatment kit could make bed nets far more effective in protecting babies and toddlers from the terror of malaria, they said, but this was also a prime opportunity to beat a local competitor in the bed-net business. "If you get into this, you could steal the margin on this," de Savigny said, "because the public sector is about to start promoting this product." Public sector institutions were never going to promote the baseball caps and T-shirts that A to Z was making at the time, de Savigny continued. But they would promote this product if A to Z had a better net than its competitor, Sun Flag.

Six months later, at a conference in Dar es Salaam, de Savigny could see that A to Z was keen to ramp up the business. The conference had reserved space for insecticide manufacturers and net makers, and Sun Flag hung its standard white conical net against a white wall. In came Shah and Haria with a net that looked completely different. It was a big green silky rectangular net, and it was such a hit that the African Medical Research Foundation ordered a hundred thousand of them on the spot. By 2003, A to Z was the largest producer of conventional—that is, not long-lasting—insecticide-impregnated bed nets in Africa, producing over six million nets a year.

Meanwhile, Sumitomo Chemical Company of Japan had developed long-lasting insect repellents that could significantly improve the bed nets, resulting in an insecticide-treated net that could last for at least five years. But as a chemical company, it lacked the expertise to package, market and distribute its wares in the developing world where insecticide-impregnated bed nets would do the most good. It needed a partner.

In 2002 the World Health Organization and the Roll Back Malaria Partnership approached Sumitomo, encouraging the multinational to look into manufacturing its Olyset nets in Africa. Sumitomo soon found that A to Z was Africa's biggest producer of conventional impregnated bed nets, with a full set of facilities and trained workers.

A to Z also had one other critical ingredient—proven technological know-how. Working with Olyset requires not only an understanding of the technology of textiles but also an understanding of the technology of plastics. It was a perfect match for Africa—one that is celebrated by a sign at the factory's front door: "Made in Africa—with Japanese Technology." A to Z is now Africa's largest manufacturer of long-lasting insecticide-impregnated bed nets.

"Malaria is in Africa," Haria told us, "so any problems to be solved should be here in Africa, not in the Far East or the First World countries. We've shown that things can be done in Africa. The image the First World has created for the Third World, especially here in Africa, is that nothing can be done. But this is a very good example that things *can* be done."

Haria and Shah led us through the modern factory, where a couple of hundred employees were working on twenty-eight production lines. The manufacturing process starts with polyethylene material and the Japanese pellets containing the insecticide. The gravel-like pellets are melted, pulled into a string and wound around a bobbin. Then this "yarn" is woven into sheets and cut into the right size for bed nets. The process is carefully monitored to make sure enough insecticide is impregnated into the nets to last five years—the minimum time for a long-lasting net. Workers pack the finished nets and load the boxes into company trucks that deliver them all over the country and beyond.

When we visited A to Z in 2008 and again in 2009, its bed-net business was booming. Between trips it had added another football-field-sized plant. The company sells virtually all of its bed nets to national malaria control programs, which buy them with funds from agencies such as UNICEF. Production of bed nets has escalated rapidly, from ten million a year to twelve million and, in 2010–11, to twenty million a year. A to Z expects to supply a significant chunk of the estimated African demand for eighty-five million nets per year over the next three years.

Three months before our visit, Shah had entertained President George Bush and his wife, Laura, a visit that was commemorated with a photo of the presidential couple that hung in a prominent position in the company's conference room. The Bushes had spent an hour at the factory, including forty minutes on the factory floor. The president had asked many questions, but Shah's favourite was, "What do you do for the ladies?" This was actually a good question since most of the workforce are women. The company provides housing, meals and medical care to its six thousand workers, in addition to wages that are much higher than those in the nearby farms.

"My biggest motivation," Shah told us, "is the number of lives we are saving. That's one. Number two: Africa needs jobs, and without people getting jobs this continent will never develop. The more jobs that are created gives me personal satisfaction that at least people are having an activity, they're earning out of it and they are benefiting out of it."

While A to Z relied on technology from Japan, its success illustrates how African entrepreneurs can successfully manufacture a key health product in the low-income countries that will use the product. The story also shows how supportive partnerships with both public and private players can mobilize a country's capacity to tackle its own health challenges. As A to Z's CEO observed, the commitment from public buyers to purchase the $6 nets in large quantities was crucial to the company's success.

As an African company, A to Z is also in a prime position to make sure that the nets actually get into homes to protect sleeping children. A to Z relies on a private distribution network of wholesalers and retailers in eastern and southern Africa. In Tanzania, its fleet of 150 trucks distributes the nets all over the country, and to Kenya, Uganda, Zambia, Zimbabwe, Mozambique, Burundi and Rwanda. A to Z truckers have displayed a remarkable ability to deliver nets on time to the remotest villages, according to Nick Brown, leader of the ITN Cell, which buys and distributes bed nets for Tanzania's National Malaria Control Programme. This represents a significant advantage over their international competitors, who typically deliver to national ports. It means that more children in sub-Saharan Africa are sleeping under bed nets and fewer of them are dying of malaria.

For a time, A to Z was the only manufacturer of long-lasting insecticide-impregnated nets that had been tested and approved by the WHO. The company now faces challenges from new competitors just when demand for nets in Africa is expected to plateau by 2013. In a way, this is a good thing: it means that all who need and want nets have them. But it's also forcing A to Z to think about other ways to use its technology. The company is beginning to innovate. It is thinking of how it might use nets to protect crops. It has hired a scientist, previously employed by the WHO, to lead its R&D on new applications. So the twin tracks of business and science have come together, albeit with some help from imported—not domestic— scientific talent.

Haria and Shah are rightly proud of A to Z's achievements. Yet it amazes us that a company like A to Z could operate successfully for so long without being aware of a scientist such as Wen Kilama, who has a deep knowledge of malaria in Tanzania. That got us thinking. How can African companies like A to Z tap into the knowledge of African scientists like Wen Kilama? What conditions need to be established in Africa to support a hundred companies like A to Z that use African technologies, talent and ideas instead of imported ones?

We have to find a way to overcome the barriers that currently block the flow of ideas from the lab to the village. There are three primary barriers: The culture of scientific research in Africa aims for publication in journals, not for turning ideas into products that save lives. Knowledge and skills do not flow among researchers, business, capital providers and government regulators and policy-makers. There is little venture capital to fund ideas at an early stage.

And yet we know that emerging economies can innovate and commercialize life-saving products. India and China are successfully developing an innovating culture. The Indian company Shantha, as we saw in the previous chapter, has developed a hepatitis B vaccine for less than a dollar a dose, a twentyfold reduction in cost, so millions more people can be vaccinated in poor countries, where most of the disease occurs. The company Sinovac produced the H1N1 vaccine domestically in China, and came out with the product before companies in the United States and Canada. If these emerging economies can innovate, why not African countries?

One of the creative things we've done in Canada is to bring together some of our best scientists, entrepreneurs and far-sighted investors under one roof at the MaRS Centre in Toronto in an effort to translate new ideas into products and services. The centre is right in the building where Banting and Best first injected a patient with insulin in January 1922. It has floors of research laboratories and houses scores of companies, including pharmaceutical companies, small biotechnology companies, venture capitalists, law firms and all the other players needed to take research and turn it into a product on the market. This is called convergence innovation. If it works for us, can it work in Africa?

We think it can. At their invitation we have now been working with four African countries—Ghana, Tanzania, Rwanda and Uganda—to help them develop African scientific ideas and turn them into real products and services that can save and improve lives in Africa. We've spent many long hours with our African colleagues thinking about practical ways to achieve this goal, and together we've created a new model of innovation for Africa.

Under this model, innovation in Africa would have three components. First, a product development fund would identify technologies stuck in laboratories and refine them to the point where they might interest private investors. Second, a virtual network would link scientists and business people. We hope this would show business people the scientific opportunities that up to now have been hidden in research institutions. At the same time, we hope the network would show researchers how business can help them develop and commercialize their ideas. This could persuade academics and researchers to expand their goals from publication in academic journals to the creation of products. Finally, an African country would build a physical centre, like the MaRS Centre, where scientists and business people could share equipment and office space, and share ideas around the water cooler, and incubate new companies.

After we refined this model, our next question was obvious: where would the money come from? We knew that the African Development Bank primarily financed infrastructure and was in the middle of deep thinking about strategy under the guidance of a high-level panel co-chaired by former Canadian prime minister Paul Martin. We pleaded our case to Martin one evening in his office on Parliament Hill in Ottawa, apparently with some success: the panel eventually released a report saying[2] that the bank should support "the development of national and regional centers of excellence in the health sciences and in energy and environmental technologies." We then met officials of the African Development Bank at their headquarters in Tunis. They liked the idea of convergence innovation centres and are waiting for African countries to apply for funding through their ministries of finance.

Next, we turned our attention to the absence of money to fund great African ideas for health products. We found we could not identify one single dollar of invested life sciences venture capital in sub-Saharan Africa outside South Africa. This was a shocking revelation. It meant that if you are a young scientist in Africa and you have a great idea, you could never move that idea from the laboratory to the village because you could not find a company that would finance turning the idea into a product. Is it any wonder there is a brain drain from Africa?

We did, however, find one ray of hope. The International Finance Corporation announced in 2007 that it would invest up to $1 billion in health care in Africa.[3] As it stands now, most of that money is to be invested in private health facilities, but we have tried to persuade the organization to invest some of the funds in life science research and commercialization. What Africa needs, as Milton Lore, head of the African Venture Capital Association, has pointed out, is a network of angel investors who would put early money into life sciences innovations and the development and commercialization of health products. These are high-risk investments, to be sure, but angels who have made their money elsewhere and believe strongly in the power of a country's private sector to raise their own people out of poverty should be game. Such civic-minded individuals, many of whom profited handsomely from being ahead of the crowd, were critical in Boston and San Francisco, and they will be just as critical in Accra, Nairobi, Dar es Salaam, Kampala and Kigali.

We continue to think the long-term answer to Africa's health problems lies in made-in-Africa solutions, and the more we work with our African colleagues, the more firm our belief. One of our early stops on the continent was Rwanda, a country with a sad history and hopeful future. To stand on the site where more than 250,000 people are buried, as one does at the genocide memorial in Kigali,[4] is almost indescribable. Then you hear the stories, and they are bone-chilling. Our guide at the memorial related how he had hidden in trees so the *genocidaires* would not find him. The stories of the alleged complicity of France in the genocide and the inaction of the United States and the United Nations are horrifying. Most shocking are the stories of how the clergy lured people into churches, places of worship and refuge, where they were murdered en masse.

These are sickening tales of what humans can do to other humans.

And yet, this tiny country is full of hope and inspiration. We were invited to Rwanda by Professor Romain Murenzi, then the minister in charge of science, technology, scientific research and information and communication technologies. It would be hard to find a more positive man. We had first met Murenzi in 2007 over breakfast in a hotel coffee shop in Ottawa. "I see you guys are doctors," he said. "If you want to talk to me about how to educate health care workers and build hospitals, I will introduce you to my colleague the minister of health. If, on the other hand, you want to talk about how we can make products, based on Rwandan know-how, and export them to other countries, I am your guy. Do you know that every week two planeloads of Rwandan flowers land in Amsterdam to be sold in the flower auctions there? How did we accomplish that? Using knowledge. We needed to study and improve how flowers were preserved, and we now are able to preserve them so we can export them to Amsterdam. This has been an important source of exports, jobs and foreign currency for us. Is this the sort of thing you are talking about in health?"

Well, yes, it was. In Kigali, Murenzi took us into a room where about ten young people were working away at computer terminals. This was a company that Murenzi had helped to create. Its workers took archival mechanical drawings and two-dimensional paper plans from clients in the United States and Europe and turned them into three-dimensional drawings on the computer. Here was an example of outsourcing to Rwanda, using bright African students, in the global economy. It was also an example of the kind of high-tech company Murenzi wanted to create in his country. He clearly had the backing of Rwanda's visionary president, Paul Kagame, who a year earlier had summed[5] up Africa's dilemma this way: "We in

Africa must either begin to build up our scientific and technological training capabilities or remain an impoverished appendage to the global economy."

In 2007 we started work on a convergence innovation centre for Rwanda that contained the three parts we'd identified as desirable— a product-development fund of $100,000 for three pilot projects involving silk, pyrethrum and banana fibre; a virtual network connecting scientists and business; and a physical centre like our own MaRS Centre in Toronto. On one of our trips, we passed one of the sites that Murenzi had proposed for the centre. It was being cleared by prisoners, all of them *genocidaires*, their uniforms colour-coded according to their level of involvement in the genocide. Now these prisoners were clearing the land for Rwanda's technological future. A more telling tableau of Africa's past and future, and all that is good and bad, would be difficult to find.

Another example of progress in developing partnerships between science and business can be seen in Tanzania. We started working with that country's government after a chance meeting in Kyoto with the minister of communications, science and technology, Professor Peter Msolla. Abdallah, speaking in Swahili, told him about our work in Ghana. The minister said that Tanzania had just the same problem—much of the research there was going to waste. He invited us to come to his country, and a few months later we were describing our idea of a convergence innovation centre to Msolla in his office in Dar es Salaam. For Abdallah, this was a poignant moment. He had come a long way from working in his father's butcher shop in that city to presenting a business plan for a high-technology centre to a Tanzanian cabinet minister. After so many years working abroad, here was an opportunity for Abdallah to give back to his country of birth.

We needed to determine whether Tanzania's scientists and business community were interested in working together, so we held a meeting in Dar es Salaam in early 2009. We were delighted to see a wave of enthusiastic support. More than a hundred people turned out for one of our workshops. Here we learned a startling fact: of the fourteen hundred patents based on research done in Tanzania, not a single one is owned by a Tanzanian. The Tanzanians peppered us with questions and comments. They talked about the costs of filing patents, the methods of scouting for potential ideas and products, and the current lack of information regarding access to research funds. They wanted to know how many products would be funded each year through the convergence innovation centre. They talked about the opportunities that would soon come with the rollout of Internet fibre optic cable in 2009 and the massive increase in bandwidth. (Tanzania has now developed an extensive fibre optic network that has reached all the way to the Rwandan border.) We were amazed: all the participants fully supported the project.

As of this writing, the steering committee set up by the minister has identified a promising site for the centre on the outskirts of Dar es Salaam, where the new campus of the Muhimbili University of Health and Allied Sciences is being built. Tanzania expects to ask the African Development Bank for financing to pay for construction of the building and five years of operating expenses. The Tanzanian government anticipates that if all goes well, the project will be financially self-sustaining in the third or fourth year of operation.

We are also working with the government of Kenya, homeland of our early mentor, Calestous Juma, professor at the Harvard Kennedy School of Government. As co-chair of the African Union High-Level Panel on Modern Biotechnology (of which Abdallah was a member), Juma has spoken forcefully about Africa's failure to nurture scientific innovation. "It is no secret," Juma has written, "that the history of development in Africa has been marked by the delivery of benefits in science, technology, and innovation to only a few of its citizens, without providing tools of development to the rest of the population. Without those tools," Juma noted, "Africa stays poor. The proportion of people living in absolute poverty in Africa is higher now compared to what it was in the 1980s and 1990s."

Kenya is another African country that, to its credit, has made strong attempts to commercialize innovations—but with decidedly mixed results. In 2004, a group of venture capitalists established the country's first life sciences VC fund, Bridgeworks Africa. Unlike traditional VC models, Bridgeworks was a non-profit that partnered with a single public institution, the International Centre for Insect Physiology and Ecology, or ICIPE. Since its inception in 1970, ICIPE had partnered with researchers from more than seventy countries to develop tools and strategies to manage insects for the benefit of people's health. The institute's scientists have conducted research on mosquito repellents and attractants, and also repellents and attractants for tsetse flies, which transmit sleeping sickness. By the early 2000s, the institute had eight patents related to these technologies.

The brainchild of Dr. Hans Herren, ICIPE's director from 1994 to 2005, Bridgeworks hoped to commercialize this rich research base. The patents provided a good start, and the VC fund received $1.5 million in seed funding. But it had trouble raising more capital, and it spread its investments too thinly across fifteen companies, some of which were managed by staff in Switzerland. As a result, Bridgeworks' efforts were unsuccessful. Although it had managed to identify and evaluate promising technologies, it tried to do too much with limited resources and didn't adapt its business model to the local contexts. It is inactive now, but its example offers important lessons.

Another valiant but ultimately unsuccessful effort to commercialize innovations in Kenya was funded by the Japan International Cooperation Agency and the government of Kenya in 2005. This was a commercialization unit at the Kenyan Medical Research Institute. Today, the unit is a free-standing centre, a new building with state-of-the-art equipment in the heart of Nairobi, where it's linked to KEMRI, a large research organization. Yet the commercialization unit stands idle. What went wrong?

The unit was built around a single technology, reverse passive hemagglutination for diagnostics, and a single product, called Hepcell, which tests blood for hepatitis B. Hepcell was much cheaper than the imported product, and for a while it sold well. But then the national blood transfusion policy and strategy changed towards adopting newer technologies, and the regulatory standard changed, adding the requirements for WHO prequalification. Since the Ministry of Health's diagnostics procurement was largely supported by donors (who normally purchase only WHO prequalified health products) and there was weak government policy for support of local innovations, the market for Hepcell collapsed. The lesson? Innovation needs to be built around more than one product, more than one technology, more than one research centre, and it needs a supportive policy framework. Otherwise it risks turning out like the KEMRI commercialization unit—empty and idle.

Both Bridgeworks and KEMRI are important efforts that should not deter future attempts to set up VCs to fund research and commercialize innovations in Africa. No major transformation in society happens flawlessly or easily. The people who initiated these projects are brave pioneers in a new world of innovation, and we have to admire their foresight and tenacity. The lessons we have learned from their experiences will surely help the African VCs of the future.

———

Working with our partners in Africa has been a real eye-opener for us. We have seen first hand the opportunities and the roadblocks they face in devising solutions for the health problems of Africans. Scientists in Africa can see the problems in their communities and their own families. They can see how to fix those problems in a way that works on the ground. But their ideas are too often squandered, too infrequently developed into meaningful products. We decry this waste of talent and ideas.

Yet, from what we have seen and learned, we are optimistic. We believe Africa will soon become better at directing the talent of its people into meaningful health products that help to develop the economy. We think this can happen if a network of innovation centres helps to break the impasse that keeps African people poor and dependent. Imagine a network of such centres in several African countries, all linked and serving as a "one-stop shop" for investment. Now imagine this not only for health-related innovations but also for agriculture and energy. Talent and ideas would not be wasted for want of resources, and Africans could use their ingenuity to solve their own problems. This is the way for poor countries not to be poor forever.

Why did we choose to work in Ghana, Rwanda, Tanzania and Uganda? Because these countries have stable governments and strong macroeconomic growth. These countries, and others such as Kenya and Nigeria and of course South Africa, can show the way forward for the rest of this dynamic continent.

We see a very bright future for Africa. It's true that right now in Africa science and business are like two trains on tracks that never cross. Promising technologies are stuck in laboratories. There is no capital to finance ideas, and talented people leave the continent. But we think this can change. We hope that Africa will one day become a continent where a hundred Wen Kilamas and A to Zs work together to create vibrant, knowledge-based, Africa-based enterprises that tackle the continent's health, agricultural and environmental problems. It won't be easy, but we think it is inevitable. The only question is how quickly the journey can be accelerated by enabling the right conditions for African talent to develop. In the old world, talented Africans like Abdallah and Calestous Juma rose from their poor African roots to become eminent professors. Today, from their posts in North America, they're contributing to the scientific and economic development of Africa. We hope that in the new world, talented Africans will be able to create solutions at home. When this happens, Africa will thrive.

CHAPTER EIGHT

When we began our decade-long journey, we were convinced that innovation—in particular, the extraordinary
power of science and technology—could solve some of the most
difficult health problems facing people in the world's poorer
countries. We believed that innovations emerging from the thrilling world of genomics could transform the way the diseases of the
poor are diagnosed, treated and prevented. And we believed that
when these new scientific tools were applied to such problems,
other benefits would follow: there would be opportunities to
create jobs and escape from poverty. Our journey has only deep-
ened this conviction.

We started this book by describing our personal experiences, how we met, and how we went on to develop a program that focuses on harnessing innovation to improve global health. Along the way, we've engaged with many fascinating people. We've learned how treacherous the terrain from the lab to the village can be, especially as it passes through ethics and commercial enterprise. And we have been awed by the opportunities and challenges in front of us. Now, as we draw this story to a close, we thought it appropriate to look back over the road from lab to village and take stock of what has been achieved in the past decade.

Our journey began with the sad story of how the lethal parasitic disease malaria killed Abdallah's sister Alwiya in 1997. Malaria, a long-time scourge of humankind, has been eliminated in much of the developed world. This accomplishment should give us hope that the same can soon be achieved in the developing world.

Today, about half the world's people are at risk of contracting malaria, and there are approximately 250 million new infections every year.[1] Malaria kills large numbers of children, mainly in sub-Saharan Africa.[2] But there is real hope that these statistics will change. In the last decade, the number of deaths from malaria has dropped significantly: from an estimated 985,000 in 2000 to 781,000 in 2009. Amazingly, about 45 percent of the one hundred or so countries where people are still infected with malaria have seen decreases of more than 50 percent. A few, like Turkmenistan and Morocco, have been officially declared malaria-free.[3] We have described in some of the previous chapters the many approaches being brought to bear against malaria: more effective drug combinations, insecticide-impregnated bed nets, residual indoor insecticide spraying and intermittent preventive treatment are all helping to control the disease. On average, countries in sub-Saharan Africa have bed net coverage of 42 percent, but in some countries coverage is at a much higher level.[4]

In Senegal, where it is around 80 percent, the number of malaria cases decreased by 41 percent in a single year.[5]

Anti-malarial drugs have become more accessible and less expensive, and better combinations containing artemisinin have been developed. Artemisinin itself, the current mainstay of malaria therapy, and so difficult to produce a few years ago,[6] is now being made incredibly cheaply. Jay Keasling at the University of California, Berkeley, has used synthetic chemistry and genetic engineering methods to make artemisinin in the laboratory, and over the past few years his team have steadily increased their yield.[7] This synthetic artemisinin is expected to come to market in 2012. When it does, it will make a significant difference in the cost of malaria treatment—so much so, in fact, that it will likely put out of business those criminals who are now making huge profits selling counterfeit preparations. Making artemisinin inexpensively will help save many more lives every year.

The ultimate cure for malaria, however, will be the development of an effective, affordable vaccine—and for this, too, there is hope. In a previous chapter we mentioned the RTS,S vaccine and the early clinical trials carried out on children in Africa.[8] This is the best vaccine we have at present, although it is not perfect in terms of the percentage of children who develop immunity after vaccination. Nonetheless, the effect is large enough that the vaccine has now entered Phase 3 trials in sub-Saharan Africa. Preliminary results show that the protection lasts for at least fifteen months[9] (the trials have not been going on for much longer that that, so we may eventually find that the effect lasts even longer). It is expected that this vaccine will be on the market by 2016. Other vaccines are also under development, including one by Sanaria, which uses irradiated sporozoites (an early developmental stage of the malaria parasite). It has failed to work as tested[10] but the company is hoping it may work if administered in a different way. Another approach, by Stefan Kappe and his colleagues in Seattle, which we discussed in Chapter Two, is the knocking out of selected parasite genes to make the parasite incapable of going through the full cycle once it enters the human body. The results of the early trials of this latter vaccine are not yet known. What we do know is that an effective vaccine against malaria will be a major milestone in global health.

There is also good news about another scourge that has been the focus of several chapters in this book: HIV. In 2009, a clinical trial in Thailand showed positive results with an HIV vaccine under development.[11] The vaccine regimen consisted of priming with a canarypox vector carrying three synthetic HIV genes, followed by booster inoculations with two recombinant envelope proteins from two types of HIV (clades B and E). This is the first Phase 3 trial to show that a preventive vaccine for HIV may be feasible.[12] To be sure, the effect was modest, but HIV vaccines so far have been repeated failures so any progress is of immense value. The trial subjects who showed protective immunity are now being intensively studied to try to understand the mechanisms of the protection.

One of the problems with HIV vaccines is that results in the laboratory, where antibodies can neutralize some types of human immunodeficiency viruses, haven't been duplicated in the field, where vaccines so far have failed to induce in the host such strong neutralizing antibodies. So a related recent development has been great news: scientists from the U.S. National Institute of Allergy and Infectious Disease Viral Research Center discovered two antibodies that can bind to a part of the human immunodeficiency virus and neutralize it.[13] Previously there had been antibodies that could bind to the virus, but these were able to neutralize less than half of the virus strains. Researchers found that these newly discovered antibodies, called VRC01 and VRC02, can neutralize nine out of ten strains.[14] This breakthrough during 2010 gives us hope that we can both understand the biology of HIV infection of humans, and that by reverse-engineering, so to speak, we can identify parts of the virus that interact with the human body in a way that makes the body produce those neutralizing antibodies. We can then make a better vaccine with that knowledge—a process called active immunization.

Moreover, with the discovery of these strongly neutralizing antibodies it is now also possible to seriously consider making large quantities of antibodies in the lab to be used for *passive* immunization. One of the pioneers of this promising work is another great scientist funded by the Grand Challenges in Global Health (GCGH) program we encountered in Chapter Two: David Baltimore, a Nobel Prize winner attracted to work on global health by GCGH funding. Baltimore and his colleagues at the California Institute of Technology are now intensely studying the structure of these highly neutralizing antibodies so that they can make such antibodies or, even more exciting, synthetic antibody-like proteins, in the laboratory. This work is at the cutting edge of the life sciences that have been shaped and energized by the genomics revolution we described in earlier chapters.

The good news with HIV doesn't stop there; there have been three other recent breakthroughs. First, adult circumcision. Clinical trials in sub-Saharan Africa have shown that adult male circumcision can reduce HIV transmission rates by around 60 percent.[15] As a result, African countries are now organizing circumcision campaigns, and although the number of men circumcised so far is relatively low,[16] there are efforts to increase these efforts by a factor of ten. It appears that young adults in sub-Saharan Africa actually welcome circumcision;[17] if this continues to be the case, adult male circumcision could make a big difference in HIV infection rates.

A second breakthrough has come from the work of Quarraisha Abdool Karim, Salim S. Abdool Karim and their colleagues at CAPRISA in South Africa. We described their work engaging communities in Chapter Four. Despite some criticisms of the CAPRISA 004 trial, including those from scientists who claimed that the design of the trial was faulty, these intrepid scientists in Durban, South Africa, showed convincingly in 2010 that vaginal microbicides can work. The 004 trial used a vaginal microbicide preparation, whose active ingredient is tenofovir, in women who were potentially exposed to HIV-positive male partners. When the Abdool Karims and their colleagues announced their results at the 2010 annual HIV congress in Vienna they received a standing ovation. But in keeping with the themes we have talked about in previous chapters, it will take some time yet before the formidable confirmatory, regulatory, ethical, socio-cultural, pricing and health systems challenges are overcome and this life-saving technology is made available at a scale large enough to save lives.

The third HIV breakthrough, which also occurred in 2010, is called PrEP—pre-exposure prophylaxis (in this context, using oral drugs). In Chapter Four we talked about the controversy surrounding the anti-retroviral drug tenofovir when plans were made to test it in Cambodia in 2004, and how those efforts were severely hampered because of the failure, or non-existence, of community engagement. Now several new trials are being conducted, with community engagement, in the United States and elsewhere.[18] So far, the largest global PrEP trial in men who have sex with men, a trial known as iPrEx, has demonstrated that tenofovir preparations can decrease HIV incidence in this population. Still, scientists are carefully evaluating the effects of non-adherence by some of the subjects taking the trial drugs; we need to evaluate if there will be changes towards more risky sexual behaviour.[19]

Unfortunately, another study of PrEP using oral drugs in African women was stopped early in April 2011 because the drugs were not working, and scientists are now reconciling the various findings.

On the tuberculosis front, the picture is mixed, but there is long-term hope. The latest figures from the World Health Organization show a decrease in the number of people infected, although there are still 1.7 million or more people dying from this curable disease every year;[20] indeed, the number of people dying from TB is second only to the number for HIV/AIDS. We contend that although the incidence is dropping, the change is still too slow; at these rates TB will not be eliminated in our lifetimes. Even more worrying is the fact that the number of people with multi-drug-resistant TB (MDR-TB) is increasing dramatically, with more than 440,000 new cases every year, of which only 5 percent are treated properly.[21] If more of these cases become prevalent in vulnerable communities, there could be an epidemic of untreatable or difficult-to-treat TB, which would be a global disaster. Fearing this, researchers are now on a desperate search for vaccines on the one hand, and new drugs on the other.

There have been virtually no new anti-TB drugs developed in the past fifty years[22]—until recently, when the Global Alliance for TB Drug Development (known as the TB Alliance) started working on this issue with academia, research institutes and industry. There is now a healthy pipeline of more than ten new drugs.[23] And along with these potential drugs has come a completely new way of thinking about drug development.

TB drugs have to be given in combinations, usually over a period of months, to reduce the likelihood of drug resistance. Until now each individual drug was developed and tested separately; if it worked, it was added to a multi-drug combination, which itself was then tested. This inefficient, decades-long, expensive process is being challenged by a new initiative, the Critical Path to TB Drug Regimens (CPTR), which is pushing for a protocol that would allow promising drug combinations to be developed and tested together from the beginning, cutting the time required from decades to just years.[24] One of the biggest hurdles to doing this in the past was the regulatory agencies, which required old-style clinical testing, but the enlightened new leadership at the Food and Drug Administration is responding positively to the recent approach, seeing it as not only applicable to TB but, in the long run, to other diseases that need combination therapies—in particular, malaria and cancer.[25] At the end of 2010, the TB Alliance launched just such a clinical trial. The NC001 Phase 2 trial—the first of its kind—is testing a three-drug combination (with two new drugs alongside one older drug) aimed at treating both drug-sensitive and multi-drug-resistant TB. If successful this will offer a shorter, simpler, safer and cheaper option, particularly for MDR-TB.[26]

There are also important developments with TB vaccines. One vaccine, called H56, from the Statens Serum Institute in Copenhagen, combines antigens currently used in prophylactic TB vaccine candidates with an antigen that persists in late infection. The results in mice, which were published[27] in February 2011, showed the vaccine worked well for early infection and performed particularly well in the persistent, late phase of TB infection, raising real hope for tackling tuberculosis in people who have latent infections. But the biggest immediate breakthrough, reported in 2010, was the development of a rapid diagnostic test, the Cepheid GeneXpert system.[28] Until now, diagnosing TB has been notoriously difficult, expensive and time consuming. It has traditionally taken up to six weeks to get test results. The GeneXpert test cuts that to less than two hours. The test has one other very important capability: it can tell if the TB bacteria in the patient have developed resistance to the important first-line drug rifampicin. Without this knowledge, patients might be treated for months without response and at great cost. The test is still expensive (approximately $17) and currently beyond the means of most developing countries, but inevitably the price will come down.[29]

We have described advancements in the "big three" diseases—malaria, HIV and TB—as an indicator of the pace of progress and the hope that inspires those of us working in global health innovation. But hope is not limited to the "big three." In January 2010 at Davos, Bill Gates declared[30] the next decade the "Decade of Vaccines." He announced that the foundation will spend ten billion dollars over the next decade, and launched the Decade of Vaccines Collaboration; Peter is on the steering committee and co-chairs the Public and Political Support working group. A key message of Gates's 2011 annual letter was that vaccines are the best investment in public health.[31] Indeed, we now have vaccines on the market against pneumonia and diarrhea, two of the biggest killers of children; these vaccines were not widely available in developing countries when our journey began.

We also have a new vaccine against meningitis in Africa. The drive to use the vaccine began in December 2010 and will continue through 2015; it aims to save an estimated 150,000 lives in twenty-five countries from Ethiopia to Senegal. U.S. multinationals market meningitis vaccines for $80 to $100; this one, manufactured by an Indian company, is available for less than 50 cents a dose[32] and represents the first product of a public-private development partnership. Contrasting this vaccine to those developed for measles, smallpox and polio, Bill Gates remarked,[33] "All those things were created because rich people got sick. This is the first vaccine that went through the whole process where there was no rich world market, and it had to be optimized at a very low price."

The other piece of good news is that the world is much closer now to eradicating polio than when our journey began. Polio is one of those terrible and terrifying diseases that both kills and maims—at its peak around 1952, it killed or paralyzed at least 24,000 people in the United States alone, and many more in other countries.[34] The disease has been practically eradicated in the developed world by effective vaccines. It has also been largely brought under control in much of the rest of the world. In 1988, while there were still about 350,000 children worldwide being killed or paralyzed by polio, the global health community adopted the goal of worldwide polio eradication—which would make polio, after smallpox, only the second disease in the world eradicated by human ingenuity and intervention.[35] Since then, a sustained vaccination campaign covering children around the world has brought down the number of infections to 1,604[36]—a remarkable achievement. However, the virus is still circulating and potentially could flare up wherever there are unvaccinated children or adults anywhere in the world. Currently there are pockets of infection remaining in just a few countries: India, Nigeria, Pakistan and Afghanistan.[37]

The continuing vaccination campaign needs about one billion dollars a year to complete—hopefully by 2015. As Bill Gates notes[38] in his 2011 annual letter, Rotary International has been an amazing fundraiser for polio, and several governments, including those of India, the U.S., the U.K. and Japan, contribute to this funding, with the Gates Foundation providing $200 million a year. This will still leave a gap of about $700 million for 2011–12. We contend that polio eradication really must be completed in the next few years, for the potential benefits to humanity are enormous and include not only lives saved but huge cost savings ($50 billion in the next twenty-five years). But the task is going to be formidable. Again, as Gates notes[39] in his annual letter, "To win these important fights, partnerships, money, science, politics, and delivery in developing countries have to come together on a global scale"—the very themes of this book.

On other fronts, progress is being made in vaccine research and development against viruses that cause cancer. As we learned in Chapter Three, non-communicable diseases will one day displace communicable diseases as the key health threat to the lives of the poor. We heard about the unconscionable delay in bringing a hepatitis B vaccine, which prevents liver cancer, to the market. Two human papilloma virus (HPV) vaccines to protect against cervical and other types of cancer are commercially available, and although a few questions—for example, the full duration of protection—still need to be answered, these two vaccines are now widely available for the rich and are increasingly being introduced into poor countries through GAVI. The challenge now is to find ways to bring that cost down, perhaps through scientific and manufacturing innovations by scientists from the developing world themselves.

This is just a sampling of progress over the past decade since our journey began. But what does the future hold?

Francis Collins, whom we featured so prominently in Chapter One in relation to the Human Genome Project, wrote[40] a book called *The Language of Life: DNA and the Revolution in Personalized Medicine*, which Abdallah reviewed[41] in *Nature* in January 2010. In that book Collins says that the era of personalized medicine is here, that there are virtually no conditions for which heredity does not play some role, and that it is nearly time for everyone to test their DNA as there may be potential information there that is of "considerable impact." We believe this is nearly true, but for much of the developing world the focus is likely to be not on individual personalized medicine but on public health applications of genomics, applying knowledge of genomes of human and other species—including pathogens like malaria and their vectors, such as mosquitoes—to develop better drugs, vaccines and diagnostics.

Looking ahead at the next ten years, we believe that genomics and other, related life sciences will enable advances in global health, in tandem with progress in addressing the many social determinants of health. In its first call for research proposals, the Global Alliance for Chronic Diseases (GACD), which we described in Chapter Three, has promised to put some $25 million in the spring of 2011 towards funding research to address hypertension in low- and middle-income countries and in underserved aboriginal populations in high-income countries.[42] We believe that such implementation research will become important in the future as we gather deeper insights into disease risk and biology. Implementation research will also be a key part of taking these scientific insights from the lab to the village.

While infectious diseases will continue to attract international attention, and we will continue to be vigilant in spotting emerging infectious diseases such as SARS and bird flu (we believe, for example, that the next HIV is probably being incubated somewhere in the world at this very moment), without a doubt chronic non-communicable diseases will become increasingly the focus of attention. Research and health resources will increasingly be allocated according to burden of disease—in contrast to our current situation, where finding the resources to apply to chronic diseases is largely neglected in most of the developing world.

Two new areas of focus in chronic non-communicable diseases in low- and middle-income countries (LMICs) will be mental health and cancer. Mental health is under-resourced even in high-income countries, and in LMICs it is simply not catered to in any significant way. Our research project to identify the grand challenges in global mental health came to an end in February 2011 and we will shortly announce the priority challenges. We and our partners in this enormous study envisage a much stronger social movement around global mental health, and hopefully the creation of a pool of research funders who would come together to address the priorities identified in this study, in the same way that chronic non-communicable diseases were addressed with the creation of GACD.

Cancer is also almost totally neglected in much of the poor world, despite being a leading and growing cause of death. In poor countries, diagnosis, therapy (including drugs and radiotherapy), rehabilitation and even pain relief and palliative care are unavailable for all but the wealthy and powerful. Paul Farmer at Partners in Health, Julio Frenk at Harvard School of Public Health, the Dana-Farber Cancer Center and others have started an important and much needed initiative called the Global Task Force on Expanded Access to Cancer Care and Control in Developing Countries—arguing that much greater resources and talent should be brought to bear on treating and preventing cancer in developing countries.[43] We believe this initiative will make a huge contribution to cancer care for the poor.

We have little doubt that these areas will grow rapidly as more clinical trials continue to migrate to, or originate in, developing countries. We believe that the time it takes to get great ideas from lab to village can and must be reduced—lives are lost while bureaucracy, inefficient regulatory regimes and other unnecessary obstacles slow down clinical trials everywhere in the world. Speeding up this process will need even better application of ethical standards to protect human subjects of research, and will require much more community engagement, as we described in Chapter Four.

We also believe the private sector will play an increasingly important role in the delivery of health care: the problems of global health are too great to be solved by any one sector alone. But in order to contribute effectively, the private sector will need to really internalize what social entrepreneurship means and not see "corporate social responsibility" simply as a means to burnish corporate reputations. The global health community will need to learn how to work with the private sector, and a key ingredient will be trust. Trust-building is difficult, and the trust created will always be fragile and open to abuse. This is why we think that trust-building mechanisms such as the social audit model we have developed and described in Chapter Five will become increasingly important. And we need to go beyond building trust to explore ways of sustainable trust maintenance.

There is also little doubt that we will see more innovation from the emerging economies of China, India, Brazil, South Africa, Mexico, Russia and others. If the current revolutions in Tunisia and Egypt, which have successfully deposed dictators, are followed by others, then the Middle East may be on the verge of producing the kinds of enormous political and human rights changes that would unlock even more innovative ideas.

Africa, too, is changing. We have recently seen how economies in Africa are growing regularly, at respectable rates each year. At a time when food prices are at their highest historical levels, our colleague Calestous Juma has just published[44] an important book, *New Harvest: Agricultural Innovation in Africa*, showing how African agriculture can be revolutionized and how the continent could become the breadbasket for the rest of the world—if it successfully harnesses innovation. That revolution will likely require modern biotechnology, including select genetically modified crops, as we described earlier in this book. African Union governments seem serious about investing in science and technology as tools for development. We hope our work, described in Chapter Seven, in helping African governments build innovation centres will continue to be supported and that one day we will see an Africa-wide network of such centres.

Finally, innovation in the developing world will increasingly flow to rich countries, turning the traditional notion of technology transfer on its head. This trend has been called "reverse innovation" by GE chairman and CEO Jeff Immelt.[45] We have emphasized elsewhere in this book the talent and ideas we've seen for ourselves in the developing world. Scientists and entrepreneurs in these countries are innovating against constraints on affordability and value for money that do not exist in the rich world. Meanwhile, in the rich world we do have huge financial problems related to the rising costs of health care. For example, astonishingly, just one organization, the Aravind Eye Care System in India, does more than half the number of cataract surgeries that are performed each year in the U.K., at equivalent clinical outcomes but at 1 percent of the cost.[46] Our rich world problems will eventually, inevitably meet solutions from the developing world. When that happens, those of us in the rich world may be key beneficiaries of global health innovation.

In closing, we want to describe our own personal "next chapter" and show how it is unfolding. To set the scene: in Chapters Two and Three we described how we had become involved in two "grand challenges" efforts, namely the Grand Challenges in Global Health and the Grand Challenges in Chronic Non-Communicable Diseases. The first led to the Bill & Melinda Gates Foundation investing about half a billion dollars (with the Canadian Institutes of Health Research and the Wellcome Trust providing smaller amounts) in discovery research. The second led to the creation of the Global Alliance for Chronic Diseases, whose members account for about 80 percent of all health and biomedical research funding available globally. We were very closely associated with the research needed to identify the grand challenges in both cases, and with the implementation of the subsequent programs.

On February 28, 2008, Peter's phone rang with news that would launch a new chapter in our journey. The caller was a bright, young and innovative official from the Canadian federal finance department. His first line after introducing himself was, "We have been reading your editorials." He had read two newspaper columns from 2005 in which, based on our experience with the grand challenges initiatives, Peter argued that Canada should change the way it gives money to the poor world by creating its own Canadian Grand Challenges program. Such a program could target, for example, non-communicable diseases such as diabetes and tackle the serious problem of access to clean water. The idea had inspired people in the finance department to consider a new way of handling foreign aid, the official said. So would Peter speak that day with the assistant deputy minister while the minister was delivering his budget address in Canada's House of Commons?

The call was intriguing: Canada was introducing in the budget an initial commitment over two years of $50 million of its foreign aid budget to a new Development Innovation Fund. This new fund would, as the finance minister put it, "support the best minds in the world as they search for breakthroughs in global health and other areas that have the potential to bring about enduring changes in the lives of millions of people in poor countries." The idea was to invest in innovative technologies like vaccines that don't require refrigeration and drought-resistant crops that could help prevent famine.

Now it was clear why the official was calling us. Canada had been inspired by the Gates Foundation's Grand Challenges program. The finance officials knew we had advised and worked with the Gates Foundation in helping to identify its Grand Challenges in Global

Health, had helped them launch their program, had served as ethics advisers to the program, that Peter was a member of their Scientific Advisory Board, and that we had received a large research grant from that program. They also knew that Abdallah had led a similar effort to identify the grand challenges in chronic non-communicable diseases, the world's biggest killers, and was working to help create a global alliance of research funding agencies around the world.

Now, for the first time in a generation, Canada was about to shake up the way that a small part of its foreign aid was delivered. It wanted to look for innovations that could solve major problems to improve the health of the poor. A later editorial[47] in the *Globe and Mail* urged Canada to "Embrace this Grand Challenge." It read: "Grand Challenges Canada is a way to do foreign aid that should be watched closely and, if it succeeds, be widely imitated. It is a bit of the Own the Podium spirit—Canada setting out to solve the world's big problems . . . Rome wasn't built in a day, and few believed Canada could own the podium in Vancouver. Grand Challenges is a model that may prove to be worth emulating—at home, too."

But first, the federal finance department asked for our help. How would Canada deliver the program? Although the $50 million fund over two years was just a drop in the bucket of Canada's annual $5 billion foreign aid budget, we could see that Canada was opening the door to a bigger opportunity to realize the mission of global health innovation to improve health in the developing world. At the same time Canada would be introducing a revolutionary innovation in policy: linking foreign aid to innovation using a grand challenges approach. In fact, Canada would be the first country in the world to apply a grand challenges approach to its foreign aid. Now we could potentially change the way at least part of the foreign aid budget is delivered—by encouraging innovation in poor countries rather than only offering a salve for their misery. This might help them stop being poor—by creating new enterprises like A to Z Textiles, which we learned about in Chapter Seven, which manufactures more than 20 million long-lasting insecticide-treated bed nets a year and employs more than six thousand people—and along the way improve health in a long-lasting way.[48]

After that call from the federal finance department in February 2008, we put in a proposal to become the implementing arm of the Development Innovation Fund for the area of global health. To do that, we set up a not-for-profit organization called Grand Challenges Canada, with Peter as CEO and Abdallah as the Chief Science and Ethics Officer. This organization is outside of government and is being hosted at our base at the McLaughlin-Rotman Centre for Global Health. It has its own governing board chaired by Joseph L. Rotman, a well-known and widely respected Canadian philanthropist and businessman, who has also had a major role in shaping the organization. We also formed an all-star international scientific advisory board, chaired by Abdallah and including scientists from all over the world, many of whom we have met in this book—scientists like Sir John Bell, Dr. Nirmal Ganguly and others. Our partners are the International Development Research Centre (IDRC) and the Canadian Institutes of Health Research (CIHR). In time, the government agreed to make Grand Challenges Canada its implementing arm for the Development Innovation Fund.

We spent eighteen months working on how Canada and Grand Challenges Canada could contribute most effectively to the vision of the Development Innovation Fund. Our conclusions, we believe, mark a revolutionary step in the world of foreign aid. At Grand Challenges Canada we are proposing a significant new way to handle the money that rich countries give to the poor. We want to empower scientists in developing countries so that they can not only solve their own problems but help their countries grow into richer and healthier places to live. This will stop poor countries from being poor. Can you imagine the impact if all G20 countries did the same thing?

We developed a broader definition of a grand challenge for this purpose: "A grand challenge is one or more specific critical barrier(s) that, if removed, would help solve an important health problem in the developing world with a high likelihood of global impact through widespread implementation."

Grand Challenges Canada was launched on May 3, 2010, at the MaRS Centre in Toronto. It was a thrilling day. The room was packed with notables from business, science and government. Canada's finance minister, James Flaherty, announced at the launch ceremony that Canada would invest $225 million in Grand Challenges Canada over five years. Some of the scientists in the room joined our international scientific advisory board, which will help us choose the challenges and projects that we will fund. Particularly moving was the speech of Dr. Mwele Malecela, a member of our scientific advisory board and now director general of the National Institutes of Medical Research in Tanzania. She spoke eloquently of the health challenges in her country, and the contributions scientists in her country could make. Also present were some of our partners, from both inside and outside of government, who will help us accomplish our goals— partners such as the Gates Foundation (represented by Dr. Carol Dahl); IDRC, a forty-year-old Canadian Crown corporation that works with researchers in developing countries to build healthier, more equitable societies, represented by its president, David Malone; and CIHR, Canada's health research organization, represented by its president, Dr. Alain Beaudet.

Our experience with the Gates Foundation's Grand Challenges

initiative taught us that in Grand Challenges Canada we needed to do several things differently. First, scientists from the developing world needed to be given the opportunity to play a much bigger role. The Gates Foundation hadn't explicitly considered where the scientists tackling the Grand Challenges came from. As it turned out, all but one of the principal investigators came from the developed world. This wasn't surprising; the scientific research funded by the Gates Foundation required elaborate scientific expertise and technology that wasn't available in poor countries. (About one-third of the scientists working on the teams came from the developing world, but only one of forty-four principal investigators came from the developing world, and he was from China.)

Grand Challenges Canada would be different. This time, researchers from developing countries would be thrust onto centre stage as the leaders of the research projects. This made perfect sense to us: these global scientists have the insights, skills and abilities to define and solve the challenges they face.

Then we took a step further: why not enable these scientists to work with their Canadian counterparts to tackle health challenges and contribute to lasting solutions? In fact, we could create a consortium of world-leading Canadian and global scientists, as well as research organizations and leaders from the business sector, to develop breakthrough solutions to global health challenges. We'd make sure, through access plans, that those solutions were available, on an affordable basis, to people who needed them most.

As we thought about how to structure Grand Challenges Canada, we made a second key decision. The Gates Foundation's Grand Challenges initiative had originally targeted discovery science. From the outset, its interest was primarily scientific invention—not business or social innovation. That was the way it was designed.

Once again, we chose a different approach. In our work with the Gates Foundation we discovered that the path from lab to village may begin with, but certainly does not end with, discovery science alone. We have learned that although scientific innovation is necessary, it is not the complete answer. In fact, as we have seen, a scientific innovation can be derailed by cultural, ethical and social issues, especially when it's time to test a new drug, vaccine or device in a community that needs it. We have also seen how important it is to manufacture a health product at an affordable cost, and how important it is to get the product to market and to the homes where it can improve health. This often requires business innovation, and these kinds of innovations—which have, for example, allowed entrepreneurs in India to slash the cost of vital medications—save many lives.

Reflecting on the Grand Challenges in Global Health, Bill Gates himself came to the same realization. In the December 20, 2010, edition of the *New York Times*, he is quoted as saying:[49] "We were naive when we began . . . back then I thought: 'Wow—we'll have a bunch of thermostable vaccines by 2010.' But we're not even close to that. I'd be surprised if we have even one by 2015." Gates went on to suggest that it could take ten or more years to move from new discovery science to real world impact.

Building on these insights, we have specifically carved out a niche in Grand Challenges Canada that emphasizes "integrated innovation"—combining science and technology innovation with social and business innovation. Social innovation (such as delivery models in health systems and paying attention to social determinants of health) and business innovation (including new business models for serving the poor and new ways to produce affordable health products) would help bring to scale the science and technology, and also make the innovations sustainable. Taken together, these two key differences—empowering developing world scientists to solve their own problems, and integrating social and business concerns with science and technology innovation—are intended to break the chain of dependency on foreign aid. We believe that integrated innovation will accelerate innovation overall, will allow innovations to have bigger effects earlier in the process, will enable scaling of the technologies, and will foster sustainability.

Our novel approach—supporting scientists in poor countries and fostering innovations that integrate science, business and social concerns—came through in the first Grand Challenges call for proposals we announced: to create technologies that can help to diagnose multiple conditions or pathogens at point-of-care. Point-of-care in a poor country can mean a village with no electricity that is several days' walk from the nearest clinic. The lack of diagnostic tools in these regions is a critical public health issue. One reason for the failure of available technologies to avert more child deaths in the developing world is the lack of effective diagnostic services. Diagnostic tools are not available to provide a simple and accurate diagnosis of the disease, plus an individual's health status, the risks of various illnesses the patient faces and the treatment options available. For malaria alone, a point-of-care test has been estimated to save 100,000 lives a year and prevent 350 million unnecessary treatments that not only drive up the cost of drugs but also the resistance to drugs.[50] If an affordable and portable test could be developed, it could distinguish the kind of malaria that is resistant to commonly used drugs from one that isn't. If a diagnostic device were simple enough, a minimally trained community worker could use it to detect disease in a remote village. Had one existed in the mid-1990s, it might have saved Abdallah's sister from an unnecessary death.

In our first funding initiative, we announced that Grand Challenges Canada would provide up to $12 million to researchers in low- and middle-income countries for five key topics that are crucial for point-of-care diagnostics. We are working with the Bill & Melinda Gates Foundation, which will invest an additional $30 million in the initiative, although its grants will fund researchers from anywhere. Together, for example, we are funding research into the collection and preparation of samples, such as saliva and blood. We are looking for simpler and cheaper ways to amplify and detect molecules, as well as new methods to detect biological signals such as proteins and other biomarkers. We're also investing in what is referred to as implementation science or implementation research: studying all the factors that would ensure the technology actually reaches those who need it most to save lives. This would include also looking at enabling technologies such as cellphones, waste management and power management that form the crucial infrastructure for diagnostic tools, especially in poor countries.

Our overall strategy is to set standards for these different components to interoperate—like the standards that permit plug-and-play in the world of computers—thereby making each component better and the diagnostic platform more affordable and flexible. This would enable a world where many tests could be used at the same time to diagnose the cause of fever in a child almost immediately and at the point-of-care.

Here is a great example of the integration of science, business and social innovation. Point-of-care diagnostics are not just a scientific invention—although they do require a lot of science and engineering. They're also a social innovation because they completely change the delivery model. They're like the cellphone of global health. Making these diagnostic tools affordable requires an innovative business approach as well, one that ideally taps into the entrepreneurial spirit of people in affected countries themselves.

This was only the first step. Grand Challenges Canada will be launching five Grand Challenges over the next five years. On March 9, 2011, in Washington, D.C., we launched a Grand Challenge on Saving Lives at Birth. Working with partners including USAID, the Gates Foundation, the government of Norway and the World Bank, we will fund innovations in science and technology, service delivery and patient demand to save lives around the time of birth. One shocking fact is that in the seventy-two-hour period between the onset of labour and two days after delivery, 150,000 women and 1.6 million children die.[51] This is precisely where we are falling behind on two of the three Millennium Development Goals on health. We need to do things differently and better—which is another way of speaking about innovation. Speaking at the launch, Hillary Clinton said it was "a real delight to be working with our Canadian friends,"[52] and Melinda Gates said, "I want to recognize in particular—Grand Challenges Canada—they have really pioneered the idea of integrated innovation—they have really pushed all of our thinking in this particular area."[53]

In addition to saving the lives of children, we want to ensure that those lives reach their full potential. Another shocking fact is that about 200 million children in the world have had their brain development impaired by the time they reach the age of five years.[54] The causes include malnutrition, infection, parenting practices and events around the time of birth. The consequences include lower wages, lower productivity and the derogation of human capital. Talk about a great way to lock a generation of children into poverty. So we will launch an initiative, which we call Saving Brains, to gain greater understanding of this long-neglected problem and develop solutions.

We will also be launching grand challenges in non-communicable disease, including possibly wellness, high blood pressure, cancer and mental health in the developing world—priorities for the latter have been developed in yet another grand challenges study, the Grand Challenges in Global Mental Health, which was spearheaded from our centre by Abdallah, together with the Global Alliance for Chronic Diseases, which Abdallah chairs, the U.S. NIH National Institute of Mental Health, the London School of Hygiene and Tropical Medicine, and the Wellcome Trust. Our approach on each of the Grand Challenges will be the same—invest in researchers in poor and middle-income countries and integrate science, business and social innovations.

Grand Challenges Canada, in its short life, has attracted much positive comment. In Seattle, at the annual GCGH meeting, Bill Gates identified Grand Challenges Canada as one of the important successful outcomes of their own Grand Challenges in Global Health initiative. We are now working with the Gates Foundation to advocate that other rich countries follow the pioneering example of Canada to adopt a similar approach to global health innovation. And of course the approach also extends to other areas, such as agriculture, environment and energy. What an enormous injection of energy that will result in, unleashing huge locked-up pools of ingenuity in developing countries, and empowering their citizens in more sustainable ways than through charity alone.

Over the past few years, an increasing number of global commentators and academics have suggested that the kinds of global challenges that we are facing (in health and other sectors) are too complex to be addressed through traditional global governance tools. They argue that new forms of global governance will be necessary to enable the deployment of coalitions of public, private and not-for-profit organizations and agencies to develop and implement the kinds of solutions that can overcome these complex, multifaceted challenges.[55]

Jean-François Rischard, former vice-president for Europe for the World Bank Group, has argued, for instance, that the dual forces of the demographic explosion and globalization have led to a "governance gap" between the power and complexity of global public issues and the power of traditional organizations to address these challenges. He suggests the creation of what he calls global issues networks to deal with the twenty most pressing issues.[56] Similarly Anne Marie Slaughter has argued for the creation of networks that cross sectoral lines as a response to the increasing complexity of global challenges.[57]

The question can be asked, however, whether the tools exist to organize and enable the appropriate networks to address these challenges. One of the most significant benefits of and opportunities presented by the grand challenges approach is that it provides a ready-made platform to enable cross-sector and multi-national contributions to address a single, focused global challenge. The grand challenges approach represents an "operating system" for different groups who want to work together towards a common and important goal. Dr. Calestous Juma at Harvard University has argued for the importance of science and innovation as critical tools for global "science diplomacy."[58] The grand challenges approach provides a framework—an operating system—through which to organize and coalesce interest and investment in critical challenges.

In terms of a concrete example of this science diplomacy, the first challenge being undertaken by Grand Challenges Canada on point-of-care diagnostics has brought together a private U.S. foundation in partnership with a Canadian not-for-profit organization. The Grand Challenge on Saving Lives at Birth has brought together organizations funded by foreign aid in three countries (Grand Challenges

Canada, USAID, and Norway's Ministry of Foreign Affairs) with a private foundation (Gates) and a multilateral institution (World Bank). Similar patterns might also emerge around other grand challenges, and the grand challenges platform could be extremely useful to help G20 countries collaborate using innovation to solve the world's greatest challenges.

This approach need not be confined to health. Imagine a world where rich and newly rich countries collaborate with Africans to address some of the deepest problems facing that continent—not only health but energy and water, agriculture and environment. Imagine a world where scientific talent in the North and South work together in a way that creates jobs and prosperity in developing countries. This would no longer be a relationship of the dependent gratefully receiving money from the rich donor. Instead, scientists would collaborate to address a noble challenge, the improvement of health in the developing world. Speaking in the language of science, which we think will increasingly become the language of diplomacy, they would find practical solutions to the problems of preventable disease, which is robbing families and entire countries of millions of children and adults.

The grand challenges approach also represents a new approach to foreign aid. As scientists, we found ourselves participating in a very heated global discussion about how the richest countries in the world should help the poorest ones. We jumped into this argument whenever possible because we're convinced it deeply affects the health of the individuals in poor countries. If foreign aid continues to act like a welfare handout, it may save lives in the short term but it will never help poor countries join the international middle class. If, on the other hand, we stop giving aid to poor countries, as one prominent African-born economist suggests, the lives lost could be incalculable. Aid is a subject that is so easily politicized and affected by ideological thinking, as if we have the luxury to think in black-and-white in a world permeated with grey complexities. We argue that aid is necessary, that it has been used inefficiently and often corruptly, but that there are creative ways of making it more effective, with more sustainable outcomes.

Clearly, countries in the developing world today are at different stages in their development process. Some, like Afghanistan, are sitting on the bottom rung, in the direst poverty, and need help to fund basic health, education and infrastructure to put them into the position where they can help themselves. Some countries need to absorb knowledge. By absorbing and copying scientific ideas from elsewhere, they can take the first steps towards an innovative culture.

One of the biggest figures in the aid debate is Jeffrey Sachs, an economic adviser to governments around the world. His book,[59] *The End of Poverty*, with a foreword from Bono, makes the case that extreme poverty can be ended in our time—by 2025. We possess, as he puts it, an "awesome power" to end the "massive suffering of the extreme poor." If we did so, we would end the instability of poor nations and "make our lives safer in the process." Yet the poorest of the poor cannot be expected to join the middle-class nations on their own, Sachs says: "Rather, it is our task to help them onto the ladder of development, at least to gain a foothold on the bottom rung from which they can then proceed to climb on their own."[60] When those are in place, "markets are powerful engines of development. Without those preconditions, markets can cruelly bypass large parts of the world, leaving them impoverished and suffering without respite."[61]

Sachs's prescription requires a massive infusion of money from rich countries. Sachs's remedy to the problem of poverty involves the biggest and most powerful global institutions, like the World Bank, the IMF and the United Nations, along with the world's richest countries, especially the U.S. He says they need to donate more, billions more.[62] This is true; we wish all countries would meet their commitments. Yet even if that happens, the problem will not go away. It will be reduced, but it will be the same. What's more, you have to wonder whether there's enough political support in rich countries to increase money to the poorest countries.

One of the most important questions is this: does charity help? One of the most provocative answers comes from a Zambian-born economist who was educated at Harvard and Oxford and then employed by Goldman Sachs and the World Bank. In her 2009 book[63] *Dead Aid*, Dambisa Moyo notes that in the past fifty years, more than $1 trillion in development-related aid has been transferred from rich countries to Africa. Has it helped the lives of Africans? No, she says. On the contrary, they are worse off. Aid, she argues, traps the countries that accept it in a form of welfare prison. They become dependent on aid. It distorts their markets, and leads to more poverty. To Moyo, the answer is to end aid, and end it fast.

Moyo, we suspect, is going to an extreme to make her point. We think aid is still necessary to deal with the pressing needs of poor countries. Some of the aid has paid for building universities, delivering health care, improving primary education, empowering women and addressing human rights abuses. It's needed to save lives in a disaster, and as a short-term way to pay for vital health needs. No one faced with a calamity that affects hundreds of thousands of lives could say, "Let's stop handing out money in a charitable way."

Bill Gates opens his 2011 annual letter by noting the tendency in these harsh economic times to cut aid budgets, pointing out that most people don't have a clear image of the benefits that aid actually provides, since aid covers many areas and was often provided for political ends rather than for its developmental impact. He believes the picture has changed and that much aid now is spent on hugely beneficial programs that improve people's lives in both the near and long term; he praises British prime minister David Cameron's leadership in growing and spending on development aid despite massive cuts elsewhere in the U.K. budget.[64]

Other countries in what we might consider the developing world are now flexing their scientific muscles, becoming the first entries into what CNN host and *Time* Editor-at-Large Fareed Zakaria calls "the post-American world."[65] We've seen how Indian and Chinese scientists and entrepreneurs are not only investing in new scientific solutions but also in new business models that are now delivering outstanding health products at a fraction of the North American cost. They don't need charity; they won't accept the relationships implicit in the charitable donation. Think of the new members of the G20, like China, India, Brazil, South Africa and South Korea. Do you really think they will adopt the charity narrative when their own growth engines are fuelled by innovation? Of course not. Rather than sticking with the charity narrative, we should think about turning these former aid recipients into donors—or rather investors—in global health innovation. By supporting innovation, we can start imagining a future where poor countries are no longer poor, where scientists in Africa and other developing countries collaborate with scientists in North America and Europe (North-South collaboration) and with scientists in emerging economies (South-South collaboration) to find solutions to the health problems of the developing world. In doing so, the scientists in the developing world may join with business to create new products for their own people—and create jobs as they do. Then we can imagine a time when they can stop asking rich countries for handouts on a routine basis. Innovation, in other words, is an exit strategy for aid.

Our idea of supporting innovation is aimed at the countries in Africa that are stable, have good governance and are growing their economies at a reasonable rate. Countries like Ghana, Tanzania and Rwanda need financial and scientific support to cultivate their scientists and bring to life their scientific ideas that have been stagnating for so long in academic journals. If we entered into new partnerships with their scientists, we could spark a scientific revolution in those countries that would have a major impact, not only on science, but on local business and ultimately on local health. Imagine what a $100 million venture capital fund investing in African life sciences ideas could accomplish.

Charity is necessary, but it cannot be the only answer to development and health challenges. We think innovation, and in particular innovation in the poor countries themselves, must be a vital part of the remedy. We know that the scientific search for solutions to the health problems of the poor will not solve the poverty trap in itself. Nonetheless, science is a significant part of the solution. If we support great scientists and entrepreneurs who are making brilliant progress in difficult circumstances—if we take the lab to the village—we will do far more than charity ever could to lift countries out of poverty. Then perhaps all children will have a chance to grow up and live healthy lives—just as our children do in North America. Ultimately, when the lab meets the village, and when the possibility of leading a healthy life is the same throughout the world, our journey will be over.

NOTES

Prologue

1   Adapted from: http://esa.un.org/unpp/index.asp
World Population Prospects, Population Database, which indicates that
approximately 94% of the world's population lives in lesser- and
least-developing nations.

2   For more detail, see: Daar, A.S. et al. (2007) Grand challenges in chronic
non-communicable diseases. *Nature* 450:494-496

3   To learn more, visit the web page of Kappe, S.H. at the Seattle Biomedical
Research Group: http://www.seattlebiomed.org/bio/kappe

Chapter One

1   Based on the following: Morogoro Urban District Population: 228,863
(2002 Census): http://www.tanzania.go.tz/2002census.pdf

2   Was built between 1979-1984, as per: http://www.tzonline.org/pdf/
ImprovementofFisheriesManagement.pdf, "... *the first plan for construction of
Mindu Dam was made in 1952 under the colonial rule. But the construction of the dam only
started in 1979 and it was completed after five years in 1984.*"

3   For estimates, see: Birbeck, G.L. (2004) Cerebral malaria. *Current Treatment Options in Neurology* 6(2):125-137; Murphy, S.C. and Breman, J.G. (2001) Gaps in the childhood malaria burden in Africa: cerebral malaria, neurological sequelae, anemia, respiratory distress, hypoglycemia, and complications of pregnancy. *Tropical Medicine and Hygiene* 64(1):57-67

4   For more detail, see: Carapetis, J. (2007) Rheumatic Heart Disease in Developing Countries. *NEJM* 5(357): 439-441, ". . . the decrease in publications reflects only the waning burden of disease among the less than 20% of the world's population living in high-income countries. For everyone else, rheumatic fever and rheumatic heart disease are bigger problems than ever. It was estimated recently that worldwide 15.6 million people have rheumatic heart disease and that there are 470,000 new cases of rheumatic fever and 233,000 deaths attributable to rheumatic fever or rheumatic heart disease each year. These are conservative estimates—the actual figures are likely to be substantially higher. Almost all these cases and deaths occur in developing countries."

5   Slater, A.F.G. (1993) Chloroquine: Mechanism of drug action and resistance in plasmodium falciparum. *Pharmac. Ther.* 57(2/3): 203-35

6   Chloroquine is considered the primary option in resource-poor regions, see: Enayati, A. and Hemingway, J. (2010) Malaria Management: Past, Present, and Future. *Annu. Rev. Entomol.* 55: 569-91

7   For more information, see WHO Malaria Report 2010 at: http://www.who.int/malaria/world_malaria_report_2010/en/index.html

8   Eradication efforts of the 1950s and 1960s were effective in industrialized, wealthy countries whereas poorer, tropical regions were not as easily addressed. To learn more, see: Bruce-Chwatt, L.J. (1987) Malaria and its control: present situation and future prospects. *Annu Rev Public Health* 8:75-110; Breman, J.G., Alilio, M.S., Mills, A. (2004) Conquering the intolerable burden of malaria: what's new, what's needed: a summary. *Am J Trop Med Hyg.* 71 (2 Suppl):1-15; Trouiller, P. et al. (2002) Drug Development for neglected

diseases: a deficient market and a public-health policy failure. *Lancet* 359:2188-94; Collins, F.H. and Paskewitz, S.M. (1995) *Annu. Rev. Entomol.* 40:195-219; and The Tropical Disease Research Progress Reports generated by the WHO/UNDP and World Bank throughout the 1990s

9   To learn more, see: Omenn, G.S. (2010) Evolution and Public Health *PNAS.* 107(1):1702-1709; and Kwiatkowski, D.P. (2005) How malaria has affected the human genome and what human genetics can teach us about malaria. *Am J Hum Genet.* 77(2): 171–92

10 The Nobel Prize in Physiology or Medicine 1907: http://nobelprize.org/ nobel_prizes/medicine/laureates/1907/index.html

11 The Nobel Prize in Physiology or Medicine 1902: http://nobelprize.org/ nobel_prizes/medicine/laureates/1902/

12 To learn more, see: McCullough, D. (1977) *The Path Between the Seas: The Creation of the Panama Canal, 1870–1914.* New York, N.Y., Simon and Schuster

13 To learn more, see: McCullough, D. (1977) *The Path Between the Seas: The Creation of the Panama Canal, 1870–1914.* New York, N.Y., Simon and Schuster

14 To learn more, see: Webb Jr., J.L.A. (2008) *Humanity's burden: a global history of malaria.* New York, N.Y., Cambridge University Press; Shah, S. (2010) *The fever: how malaria has ruled humankind for 500,000 years.* New York, N.Y., Farrar, Straus and Giroux; and Packard, R.A. (2007) *The making of a tropical disease: a short history of malaria.* Baltimore, M.D., The John Hopkins University Press

15 See endnote 8

16 UNICEF website: http://www.unicef.org/immunization/index_why.html "The parasitic disease malaria is responsible for a staggering number of deaths—over one million a year—the majority children under five. A child dies every 30 seconds from malaria, many in just days after infection."

17 Trouiller, P. et al. (2002) Drug Development for neglected diseases: a deficient market and a public-health policy failure. *Lancet* 359:2188-94

18 Essentially it was brought under control through various measures in the developed world by the 1960s/1970s. To learn more, see: Harries, A.D., Dye, C.

(2006) Tuberculosis. *Annals of Tropical Medicine & Parasitology* 100(5):415-431 and Raviglione, M.C., Pio, A. (2002) Evolution of WHO policies for tuberculosis control, 1948-2001 *Lancet* 359:775-80

19 See WHO Fact Sheet on Tuberculosis 2010/2011: http://www.who.int/tb/publications/2010/factsheet_tb_2010_rev21feb11.pdf. Also note that, "Sub-Saharan Africa has by far the highest annual incidence rate but the most populous countries of Asia hold the largest number of cases. India, China, Indonesia, Bangladesh and Pakistan together account for over half the global burden. Eighty percent of new cases live in 22 high-burden countries (HBC)." As per, Davies, P.D.O. (2003) The world-wide increase in tuberculosis: how demographic changes, HIV infection and increasing numbers in poverty are increasing tuberculosis. *Ann Med.* 35:235-243

20 Harries, A.D., Dye, C. (2006) Tuberculosis. *Annals of Tropical Medicine & Parasitology* 100(5):415-431

21 Young, D.B., Gideon, H.P., Wilkinson, R.J. (2009) Eliminating latent tuberculosis. *Trends in Microbiology* 17; 5:183-88

22 Davies, P.D.O. (2003) The world-wide increase in tuberculosis: how demographic changes, HIV infection and increasing numbers in poverty are increasing tuberculosis. *Ann Med.* 35:235-243; and Ducati, R.G. et al. (2006) The resumption of consumption—a review on tuberculosis. *Mem Inst Oswaldo Cruz* 101(7):697-714

23 See endnote 19; also in 2008, there were an estimated 9.4 (range, 8.9–9.9 million) million incident cases (equivalent to 139 cases per 100 000 population) of TB globally—see WHO Report 2009: http://www.who.int/tb/publications/global_report/2009/update/tbu_9.pdf

24 Or worse, they obtain partial treatment. This is not only ineffective but it offers useful undercover information to the TB bug. The TB bug learns from the ineffective drug treatment and develops a new defence against the drug. Then it becomes a dangerous Extremely Drug Resistant TB that causes epidemics in crowded places like South African hospitals and Russian prisons.

25 Based on: Piot et al. (2001) The Global Impact of AIDS. *Nature* 410:968-973

26 Ramiah, I. and Reich, M.R. (2005) Public-Private partnerships and Antiretroviral Drugs for HIV/AIDS: Lessons from Botswana. *Health Affairs* 24(2):545-51; Garrett, L. (2005) The lessons of HIV/AIDS. *Foreign Affairs* 84(4):51-64

27 76.9, see: http://apps.who.int/whosis/database/life_tables/life_tables_ process.cfm?path=whosis,life_tables&language=english

28 In 2008, there were some 33.4 million [31.1 million-35.8 million] people living with HIV, 2.7 million [2.4 million-3.0 million] new infections and 2 million [1.7 million-2.4 million] AIDS-related deaths. 60 million have been infected. Sourced at: http://data.unaids.org/pub/FactSheet/2009/ 20091124_FS_global_en.pdf

29 Hotez, P.J., Fenwick, A., Savioli, L., Molyneux, D.H. (2009) Rescuing the bottom billion through control of neglected tropical diseases. *Lancet* 373: 1570-75

30 Wardlaw, T.M. et al. (2006) Pneumonia: the forgotten killer of children. World Health Organization/UNICEF accessed at: http://www.who.int/ child_adolescent_health/documents/9280640489/en/index.html

31 Estimated 1.5 million: UNICEF/WHO (2009) Diarrhoea: why children are still dying and what can be done. World Health Organization/ UNICEF, accessed at: http://whqlibdoc.who.int/publications/2009/ 9789241598415_eng.pdf

32 To learn more, visit: http://www.globalforumhealth.org/About/10-90-gap

33 See endnote 17

34 Blumberg, B.S. (1977) *Science* 197:17-25; Blumberg, B.S. Hepatitis B Virus, the vaccine, and the control of primary cancer of the liver. *PNAS* 94(14):7121-25

35 To learn more, visit: http://www.ornl.gov/sci/techresources/Human_ Genome/home.shtml

36 To learn more, read: Jarcho, S. (1993) *Quinine's predecessor: Francesco Torti and the early history of cinchona.* Baltimore, M.D., The John Hopkins University Press; and Rocco, F. (2003) *The miraculous fever-tree.* London, U.K., Harper Collins Publishers

37 See endnote 36 and Woodward, R., Doering W. (1944) The Total Synthesis of Quinine. J Am Chem Soc. 66:849; Kaufman, T.S., Rúveda, E.A. (2005) Die Jagd auf Chinin: Etappenerfolge und Gesamtsiege. Angewandte Chemie, Int. Ed. 117(6): 876–907.

38 Daar, A.S., Mattei, J.-F. (1999) Chapter 6: The Human Genome Diversity Project. Medical Genetics and Biotechnology: Implications for Public Health. December 1999, World Health Organization

39 Jomaa, H. et al. (1999) Inhibitors of the nonmevalonate pathway of isoprenoid biosynthesis as antimalarial drugs. Science 285:1573-76

40 Schlitzer, M. (2007) Malaria Chemotherapeutics Part I: History of antimalarial drug development, currently used therapeutics, and drugs in clinical development. ChemMedChem 2:944-986

41 See endnote 40

42 See http://www.nytimes.com/2003/09/30/science/conversation-with-eva-harris-simple-side-high-tech-makes-developing-world-better.html?pagewanted=1

43 See: Carr, K. (1999) Cuban biotechnology treads a lonely path. Nature 398(6726 suppl.):A22-A23; Singer, P.A., Daar, A.S. (2001) Harnessing Genomics and Biotechnology to Improve Global Health Equity. Science 294:87-89; http://www.pugwash.org/reports/ees/ees8d.htm; Jodar, L., Feavers, I.M., Salisbury, D., Granoff, D.M. (2002) Development of vaccines against meningococcal disease. The Lancet 359:1499-508

44 http://www.icgeb.trieste.it/mammalian-biology-malaria.html; http://www.malariavaccine.org/files/010711-MVI-India-pr.htm

45 Singer, P.A., Daar, A.S. (2001) Harnessing Genomics and Biotechnology to Improve Global Health Equity. Science 294:87-89

46 Daar, A.S. et al. (2002) Top ten biotechnologies for improving health in developing countries. Nature genetics 32:229-232

47 http://partners.nytimes.com/library/national/science/062700sci-genome-text.html

48 Collins, F.C. et al. (2003) A vision for the future of genomics research. *Nature* 422:835-847

49 See http://www.unac.org/en/link_learn/canada/pearson/speechgollancz.asp

50 See http://www.smithsonianeducation.org/migrations/zoofood/rosper.html

Chapter Two

1 Hilbert, D., *Bull. Am. Math. Soc.* 8, 437: 1901–02

2 Daar, A.S. et al. (2002) Top ten biotechnologies for improving health in developing countries. *Nature genetics* 32:229-232

3 For more information, see: Dalkey, N.C. (1972) The Delphi method: an experimental application of group opinion. In Dalkey, N.C., Rourke, D. L., Lewis, R., & Snyder, D. (Eds.) *Studies in the quality of life.* Lexington, M.A.: Lexington Books; McCampbell, C., & Helmer, O. (1993) An experimental application of the Delphi method to the use of experts. *Management Science* 9(3), 458-467; Weaver, W. T. (1971) The Delphi forecasting method. Phi Delta *Kappan* 52(5), 267-271

4 $200 million grant to accelerate research on "Grand Challenges" in global health, Bill & Melinda Gates Foundation, January 26, 2003, accessed: http://www.gatesfoundation.org/press-releases/Pages/grant-for-grand-challenges-in-global-health-030126.aspx

5 See endnote 4

6 The Nobel Prize in Physiology or Medicine 1989, Nobelprize.org, accessed: http://nobelprize.org/nobel_prizes/medicine/laureates/1989/

7 Varmus, H. et al. (2003) Grand challenges in global health. *Science* 302(5644):398-399

8 For quote, see: http://www.airlie.com/about/index.htm

9 To learn more, see: Baumhover, A.H. (1966) Eradication of the screwworm fly—an agent of myiasis. *J. Am. Med. Assoc.*, 196: 240-248.; Bushland, R.C. (1985) Eradication program in the southwestern United States. *Symposium on eradication*

of the screwworm from the United States and Mexico. Misc. Pub. Entomol. Soc. Am., 62: 12-15; Knipling, E.F. (1960) The eradication of the screwworm fly. Sci. Am., 203: 4-48

10 See http://www.time.com/time/covers/0,16641,20000731,00.html

11 See endnote 7

Chapter Three

1   To learn more, read: Daar et al. (2007) Grand Challenges in Chronic non-Communicable Diseases. Nature 450(22): 494-6 and visit the site WHO Chronic Disease and Health Promotion: http://www.who.int/chp/en/

2   See endnote 1

3   To learn more: Warwick, H. and Doig, A. (2004) Smoke the Killer in the Kitchen. Indoor Air Pollution in Developing Countries. ITDG Publishing; and Bruce N. et al. (2000) Indoor Air Pollution in Developing Countries: a major environmental and public health challenge. Bulletin of the World Health Organization 78(9): 1078-92

4   "NCDs are by far the major cause of death in lower-middle, upper-middle, and high-income countries, and by 2015, they will also be the leading cause of death in low-income countries." See: Adeyi, O. et al. (2007) Public policy and the challenge of chronic noncommunicable diseases. The World Bank, http://siteresources.worldbank.org/INTPH/Resources/PublicPolicyandNCDsWorldBank2007FullReport.pdf

5   For China, see: http://www.idf.org/press-releases/idf-press-statement-china-study, and the original study, Yang, W. et al. (2010) Prevalence of Diabetes among men and women in China. NEJM 362(12):1090-1101 at http://www.nejm.org/doi/pdf/10.1056/NEJMoa0908292; For India, see: IDF Atlas at (http://www.diabetesatlas.org/content/south-east-asia): Current estimates show that 7% of the adult population, or 58.7 million people, will have diabetes in 2010.

6   See endnote 5

7 See http://www.nytimes.com/2006/09/13/world/asia/13diabetes.

8 Sakhuja, V. and Sud, K. (2003) End-stage renal disease in India and Pakistan: Burden of disease and management issues. *Kidney International* 63: S115-S118; Modi, G.K. And Jha, V. (2006) The Incidence of end-stage renal disease in India: A population-based study ESRD incidence in India. *Kidney International* 70:2131-2133

9 Ridderstrale, M. & Groop, L. (2009) Genetic dissection of type 2 diabetes. *Molecular and Cellular Endocrinology* 297: 10-17

10 Kamadjeu, R.M., Edwards, R., Atanga, J.S., Kiawi, E.C., Unwin, N., Mbanya, J.C. (2006) Anthropometry measures and prevalence of obesity in the urban adult population of Cameroon: an update from the Cameroon Burden of Diabetes Baseline Survey. *BMC Public Health* 6:228

11 Mbanya, J.C. et al. (2010) Diabetes in sub-Saharan Africa. *Lancet* 375(9733):2254-66.

12 To learn more, see: http://www.who.int/mediacentre/factsheets/fs317/en/index.html

13 See http://www.boston.com/bostonglobe/editorial_opinion/oped/articles/2009/08/17/the_next_health_frontier_chronic_diseases_in_africa/

14 7.6 Million deaths globally in 2008: Boyle, P., Levin, B., eds. (2008) World Cancer Report 2008. Lyon, France: World Health Organization, International Agency for Research on Cancer

15 See http://www.who.int/features/qa/15/en/index.html

16 Baleta A. Africa's struggle to be smoke free. (2010) *Lancet* 375 (9709): 107-8
"Developing countries will bear 60 percent of the world's cancer burden by 2020 and 70 percent by 2030, but are not prepared for the looming crisis, cancer experts warned in a report on Thursday. (. . .) There were 12.67 million new cases of cancer worldwide in 2008, with developing nations representing 56 percent of the total. By 2020, there will be an estimated 15 million new cancer cases, 60 percent of which will be in the developing world." Quoted from the following: Lyn, T.E. (2010) *Developing Nations to bear*

cancer brunt. Reuters, Thomson Reuters. Accessed at:http://mobile.reuters.com/article/healthNews/idUSTRE67I1GT20100819

17 To learn more, see: Lonn, E., Salim, Y. (2009) Polypill: the evidence and the promise. *Current Opinion in Lipidology* 20(6): 453-59

18 See http://www.who.int/mediacentre/factsheets/fs339/en/index.html

19 See http://www.who.int/fctc/signatories_parties/en/index.html

20 See endnote 20

21 WHO (2009) The WHO report on the global tobacco epidemic, 2009: implementing smoke free environments. WHO http://whqlibdoc.who.int/publications/2009/9789241563918_eng_full.pdf

22 Jha, P. and Chaloupka, F. (2000) *Tobacco Control in Developing Countries*. Oxford University Press

23 Daar et al. (2007) Grand Challenges in Chronic non-Communicable Diseases. *Nature* 450(22): 494-6

24 See endnote 24

25 Edwards, R. et al. (2000) Hypertension prevalence and care in an urban and rural area of Tanzania. *Journal of Hypertension* 18(2):145-152

26 He, J. (2005) Major causes of death among men and women in China. NEJM 353:1124-1134

27 To learn more, see: Madon, T. et al. (2007) Implementation Science. *Science* 318:1728-1729

28 To learn more, see: Research, Development and commercialization of the Kenya ceramic Jiko and other improved biomass stoves in Africa accessed at: http://www.solutions-site.org/cat2_sol60.htm

Chapter Four

1 Tsai, C.C. et al. (1995) Prevention of SIV infection in macaques by (R)–9-(2 phosphonylmethoxypropyl) adenine. *Science* 270:1197-1199

2 "Worldwide during 1995, 4.7 million new HIV infections occurred

(an average of 13,000 new infections each day). Of these, 2.5 million (an average of nearly 7,000 new infections per day) occurred in Southeast Asia and 1.9 million infections (over 5,000 new infections per day) were in sub-Saharan Africa. The industrialized world accounted for about 170,000 new HIV infections (nearly 500 new infections per day; less than 4 percent of the global total)." Taken from: Mann, J. and Tarantola, D. (1996) *AIDS in the world*. Global AIDS Policy Coalition, New York, N.Y., Oxford University Press

3   Gilead Sciences holds the following in its product portfolio: Viread® (tenofovir disoproxil fumarate) (U.S. approval, 2001; E.U. approval, 2002; U.S. and EU approval, 2008.)

4   See http://clinicaltrials.gov/ct2/show/NCT00078182

5   Cohen, J. (2003) Can a drug provide some protection? *Science* 301 (5640): 1660-1661

6   Singh, J. and Mills, E.J. (2005) The abandoned trials of pre-exposure prophylaxis for HIV: what went wrong? *PLoS Medicine* 2(9):0824-0827

7   Ahmad, K. (2004) Trial of antiretroviral for HIV prevention on hold. *Lancet Infect Dis* 4: 597.; Singh, J.A., Mills, E.J. (2005) The abandoned trials of pre-exposure prophylaxis for HIV: What went wrong? *PLoS Med* 2(9): 0824-0827

8   "Hun Sen, a former Khmer Rouge officer and the foreign minister of the PRK, was elected as his replacement." From Ayres, D.M. (2000) *Anatomy of a crisis: education, development, and the state in Cambodia, 1953-1998*, University of Hawaii Press

9   See http://www.actupny.org/reports/Bangkok/gilead.html (associated press)

10 See http://www.washingtonpost.com/wp-dyn/world/issues/bodyhunters/

11 To learn more about the controversy, see: Wise, J. (2001) Pfizer accused of testing new drug without ethical approval. *BMJ* 322(7280):194; Lenzer, J. (2006) Secret report surfaces showing that Pfizer was at fault in Nigerian drug tests. *BMJ* 332(7552):1233; Kovac, C. (2001) Nigerians to sue US drug company over meningitis treatment. *BMJ* 323(7313):592; Wollensack, A.F.

(2007) Closing the Constant Garden: The Regulation and Responsibility of US Pharmaceutical Companies Doing Research on Human Subjects in Developing Nations. *Wash. U. Global Stud. L. Rev.* 747; Jegede, A.S. (2007) What led to the Nigerian boycott of the polio vaccination campaign? *PLoS Medicine* 4(3):e73; and Nwabueze, R.N. (2004) Ethical reviews of research involving human subjects in Nigeria: legal and policy issues. *Ind. Int'l & comp. L. Rev.* 14:87116

12 See http://www.nybooks.com/articles/archives/2005/oct/06/the-body-hunters

13 Parashar, U.M. et al. (2006) Rotavirus and severe childhood diarrhea. *Emerging Infectious Diseases* 12(2):304-306

14 See http://www.contentnejmorg.zuom.info/cgi/content/full/362/4/358; http://www.stanford.edu/~siegelr/ShadmanRotashieldPaper.pdf; and Offit, P. (2002) The future of rotavirus vaccines. *Seminars in Pediatric Infectious Diseases* 13(3): 190-195

15 To learn more, see: Editorial (2005) The trials of tenofovir trials. *Lancet* 365(9465):1111; Building collaboration to advance HIV prevention: global consultation on tenofovir pre-exposure prophylaxis research (2005) International AIDS Society, Bill and Melinda Gates Foundation, U.S. National Institutes of Health, and U.S. Centers for Disease Control and Prevention; and following the trials UNAIDS conducted a series of meetings and developed the following standards for clinical trials in HIV/AIDS: http://data.unaids.org/pub/Manual/2007/jc1349_ethics_2_11_07_en.pdf

16 Berndtson K. et al. (2007) Grand challenges in global health: ethical, social, and cultural issues based on key informant perspectives. *PLoS Med* 4: e268. doi:10.1371/journal.pmed.0040268

17 World Health Organization (1997) Guidelines on the use of insecticide-treated mosquito nets for the prevention and control of malaria in Africa. Available: http://www.who.int/malaria/docs/pushba.htm

18 Mbali, M. HIV/AIDS Policy Making in Post-apartheid South Africa in: State of

the nation: South Africa, 2003-2004 (ed) Daniel, J., Habib, A. and Southall, R.; Chigwedere, P. et al. (2008) Estimating the lost benefits of antiretroviral drug use in South Africa. *Journal of Acquired Immune Deficiency Syndromes* 49(4): 410-15

19 MacQueen, K.M. and Karim, Q.A. (2007) Adolescents and HIV clinical trials. *J Assoc Nurses AIDS Care* 18(2): 78-82

20 To learn more, visit the following CAPRISA webpages: http://www.caprisa.org/joomla/index.php/researchtraining/research-sites; and http://www.caprisa.org/joomla/Micro/CAPRISA%20004%20 Backgrounder_20%20July%202010.pdf

21 See endnote 20

22 To learn more, see: Baleta, A. (2007) A second chance for microbicides. *The Lancet* 370:17; and Karim, Abdool S.S., Baxter, C. (2009) Antiretroviral prophylaxis for the prevention of HIV infection: Future implementation challenges. *HIV Therapy* 3:3-6

23 Karim, Q.A. et al. (2010) Effectiveness and safety of Tenofovir gel, an antiretroviral microbicide, for the prevention of HIV infection in women. *Science* 329:5996:1168-1174

24 Check, E. (2007) HIV trial doomed by design, say critics. *Nature* 448:110

25 Nolen, S. 28: *Stories of AIDS in Africa*. Random House of Canada Limited, Toronto, 2007

26 See endnote 25

27 See endnote 25

28 See endnote 25

29 See endnote 25

30 See endnote 23 and the CAPRISA Press Release at: http://www.caprisa.org/joomla/Micro/CAPRISA%20004%20Press%20 Release%20for%2020%20July%202010.pdf

31 Leahy, J.L.; Cefalu, W.T. (2002) *Insulin Therapy*. Taylor & Francis. Page 74

32 The, M.J. (1989) Human insulin: DNA technology's first drug. *American Journal of Hospital Pharmacy* Vol 46, Issue 11_Suppl, S9-S11

33 According to Adrian Dubock's editorial in *Nutrition Reviews*, 67(1):17-20—
"Despite the validity of current interventions, around 6000 people die
daily as a result of vitamin A deficiency as they cannot afford a suitably
varied diet."

34 See http://www.who.int/nutrition/topics/vad/en/index.html

35 West Jr., K.P. and Darnton-Hill, I. (2008) *Vitamin A Deficiency in Nutrition
and health in developing countries*. Humana Press (http://books.google.ca/books?id
=RhH6uSQy7a4C&dq=vitamin+a+deficiency&lr=&source=gbs_navlinks_s)

36 Ye, X., Al-Babili, S., Kloti, A., Zhang, J., Lucca, P., Beyer, P., Potrykus, I. (2000)
Engineering the provitamin A (beta-carotene) biosynthetic pathway into
(carotenoid-free) rice endosperm. *Science* 287:303-305

37 See: The Intellectual and Technical Property Components of pro-
Vitamin A Rice (GoldenRiceTM): A Preliminary Freedom-To-Operate Review
http://www.isaaa.org/kc/Publications/pdfs/isaaabriefs/Briefs%2020.pdf;
Potrykus, I. (2001) Golden Rice and Beyond. *Plant Physiology* March 2001,
Vol. 125, pp. 1157–1161

38 See http://www.time.com/time/covers/0,16641,20000731,00.html

39 See http://www.iphandbook.org/

40 Potrykus, I. (2010) Regulation must be revolutionized. *Nature* 466:561

41 "GM foods currently available on the international market have undergone
risk assessments and are not likely to present risks for human health any more
than their conventional counterparts. The risk-assessment guidelines specified
by CAC are thought to be adequate for the safety assessment of GM foods
currently on the international market. Guidelines for environmental risk
assessment have been developed under the Convention on Biological Diversity.
The potential risks associated with GMOs and GM foods should be assessed
on a case-by-case basis, taking into account the characteristics of the GMO or
the GM food and possible differences of the receiving environments." See:
http://www.who.int/foodsafety/publications/biotech/biotech_en.pdf

42 See http://www.gmsciencedebate.org.uk/ "The report by a panel of experts

has found no scientific case for ruling out all GM crops and their products, but nor does it give them blanket approval. It emphasises that GM is not a single homogeneous technology and its applications need to be considered on a case-by-case basis. (. . .) Worldwide there have been no verifiable ill effects reported from the extensive consumption of products from GM crops over seven years by humans and livestock. Some argue that this, combined with the testing required for regulatory clearance, provides important assurance of safety. But others argue for additional research including epidemiological surveillance. Such surveillance is very difficult for any whole food, GM or otherwise, although work is being taken forward in this area."

43 See http://www.gmfreecymru.org/open_letters/Open_letter12Feb2009.html

44 See: European Technology Assessment Group, Annex 6 Case study Transgenic crops Final Report Agricultural Technologies for Developing Countries STOA Project "Agricultural Technologies for Developing Countries," http://www.itas.fzk.de/deu/lit/2009/meye09a_annex6.pdf

45 Tang, G., Qin, J. et al. (2009) Golden Rice is an effective source of vitamin A. Am J Clin Nutr. 89(6):1776-83

46 See endnote 44

47 See endnote 46

48 See http://news.bbc.co.uk/2/hi/africa/2233839.stm

49 See http://news.bbc.co.uk/2/hi/africa/2412603.stm

50 Daar et al. Beyond GM foods: genomics, biotechnology and global health equity. in Bioethics in a small world (eds) Thield, F. and Ashcroft, R.E. *Bioethics in a small world.* Springer-Verlag, Berlin Heidelberg 2005

51 To learn more see: UNICEF Website, *Why are children dying:* http://www.unicef.org/immunization/index_why.html

52 WHO currently estimates there may be 50 million dengue infections worldwide every year. See: http://www.who.int/mediacentre/factsheets/fs117/en/index.html

53 See http://apps.who.int/tdr/publications/tdr-research-publications/
seb_topic1/pdf/seb_topic1.pdf; Dame, D.A. et al. (2009) Historical
applications of induced sterilization in field populations of mosquitoes.
Malaria Journal 8 (Suppl 2)

54 Dubbed "Terminator" technology by RAFI—now ETC Group—in 1998, see:
http://www.etcgroup.org/en/issues/terminator_traitor

55 See http://www.wired.com/science/planetearth/news/2008/01/
gm_insects

56 See http://www.petitiononline.com/NyamukGM/petition.html

57 To learn more, see: Roe, E.M. (1989) Narrative analysis for the policy
analyist: a case study of the 1980-1982 medfly controversy in California.
Jour. Pol. Ana. Mgmt. 8(2):251-273; Vreyson, M.J.B. et al. (2006) The sterile
insect technique as a component of sustainable area-wide integrated pest
management of selected horticultural insect pests. *Journal of fruit and ornamental
plant research* 14(Supp 3):107-31; Hendrichs et al. (2002) Medfly areawide
sterile insect technique programmes for prevention, suppression or eradication: the importance of mating behvior studies. *Florida Entomologist* 85(1) 9;
http://www.senasa.gob.pe/servicios/eng/students_researchers/significant_
topics/i0015-4040-085-01-0001.pdf)

Chapter Five

1   Ellen, N. (1990) Nestle infant formula controversy: restricting the marketing
practices of multinational corporations in the third world. *Transnat'l Law.*

2   Interagency Group on Breastfeeding Monitoring (IGBM). (1997) Cracking
the code. London IGBM

3   Yamey, G.: Nestlé violates international marketing code, says audit. BMJ 2000,
321:8.; Wise J. (1997) Baby milk companies accused of breaking marketing
code. BMJ. 31(7083):830-831; Newton, L.H. (1999) Truth is the daughter of
time: the real story of the Nestle Case. *Business and Society Review* 104(4):367-395

4   Singh et al. (2010) Shared principles of ethics for infant and young child
    nutrition in the developing world. BMC Public Health 10:321

5   Heywood, M. (2009) South Africa's treatment action campaign: combining
    law and social mobilization to realize the right to health. Journal of Human Rights
    Practice 1(1):14-36; Ferreira, L. (2002) Access to affordable HIV/AIDS drugs:
    The human rights obligations of multinational pharmaceutical corporations.
    Fordham L. Rev. 71:1133

6   Daar, A.S., Acharya, T., Filate, I., Thorsteinsdóttir, H., Singer, P.A. Beyond GM
    Foods: Genomics, Biotechnology and Global Health Equity. In: Thiele, F.,
    Ashcroft, R., eds. Bioethics in a Small World. Berlin: Springer Verlag, 2005

7   Based on the following: "10.9 million children under five die in developing
    countries each year. Malnutrition and hunger-related diseases cause
    60 percent of the deaths," The State of the World's Children, UNICEF, 2007;
    "Undernutrition contributes to 53 percent of the 9.7 million deaths of
    children under five each year in developing countries," Under five deaths
    by cause, UNICEF, 2006; and "each year malnutrition is implicated in about
    40% of the 11 million deaths of children under five in developing coun-
    tries," http://www.unicef.org/nutrition/index_bigpicture.html

8   Tracking Progress on Child and Maternal Nutrition. UNICEF 2009 Report,
    http://www.unicef.org/publications/files/Tracking_Progress_on_Child_
    and_Maternal_Nutrition_EN_110309.pdf

9   Trouiller, P. et al., (2002) Drug Development for neglected diseases: a
    deficient market and a public-health policy failure. Lancet 359:2188-94

10 Cohen, J. et al. (2010) Development and access to products for neglected
    diseases. PLoS ONE 5(5): 1-5

11 For detail, see: Mackie, J. et al. (2006) Corporate social responsibility
    strategies aimed at the developing world: perspectives from bioscience
    companies in the industrialized world. International Journal of Biotechnology
    8(102):103-118; Collins, K.L. (2004) Profitable gifts: a history of the Merck
    Mectizan donation program and its implications for international health.

Perspectives in Biology and Medicine 47(1):100-109; Thylefors, B., Alleman, M.M., and Twum-Danso, N.A.Y. "Operational Lessons from 20 years of the MECTIZAN Donation Program for the Control of Onchocerciais," *Tropical Medicine and International Health* 2008:13(5):689-696; and http://www.merck.com/corporate-responsibility/access/access-developing-emerging/mectizan-donation-riverblindness/approach.html

12 For more detail, see: Peters, D.H. and Phillips, T. (2004) Mectizan donation program: evaluation of a public-private partnership. *Tropical Medicine and International Health* 9(4):A4-A1; Collins, K.L. (2004) Profitable gifts: a history of the Merck Mectizan donation program and its implications for international health. *Perspectives in Biology and Medicine* 47(1):100-109; Thylefors, B., Alleman, M.M., and Twum-Danso, N.A.Y. "Operational Lessons from 20 years of the MECTIZAN Donation Program for the Control of Onchocerciais," *Tropical Medicine and International Health* 2008:13(5):689-696

13 See http://www.merck.com/corporate-responsibility/access/access-developing-emerging/mectizan-donation-riverblindness/performance.html

14 See http://www.pfizer.com/responsibility/global_health/international_trachoma_initiative.jsp

15 See http://www.who.int/blindness/causes/priority/en/index2.html

16 See endnote 15

17 Frew et al. (2009) A Business plan to help the global south in its fight against neglected diseases. *Health Affairs* 28(6):1760-1773

18 See endnote 17

19 Buse, K., Walt, G. (2000) Global public-private partnerships, part I: a new development in health? *Bull World Health Organ* 78: 549-61.

20 See http://www.iavi.org/about-IAVI/Pages/history.aspx; Chataway, J. and Smith, J.(2006) The international AIDS Vaccine Initiative (IAVI): is it getting new science and technology to the world's neglected majority? *World Development* 34(1):16-30

21 See endnote 20

22 "DNDi's malaria-specific portfolio aims to facilitate the widespread availability of the two products delivered by its diverse partners in the Fixed-Dose Artesunate Combination Therapy (FACT) Project." See http://www.dndi.org/diseases/malaria/dndi-strategy.html

23 See http://www.tballiance.org/new/portfolio.php; Ma, Z. et al. (2010) Global tuberculosis drug development pipeline: the need and the reality. *The Lancet* 375(9731):2100-2109

24 See http://www.meningvax.org/

25 Boehme, C.C. et al. (2010) Rapid molecular detection of tuberculosis and rifampin resistance. *NEJM* 363(11):1005-15

26 Cohen, J. et al. (2010) Development and access to products for neglected diseases. *PLoS ONE* 5(5): 1-5

27 See http://www.malariavaccine.org/files/FS_RTSS_FINAL.pdf; Casares, S. (2010) The RTS,S malaria vaccine. *Vaccine* 28(31):4880-4894

28 See endnote 27

29 See endnote 27

30 See http://www.malariavaccine.org/files/FACTSHEETRTS-2.pdf

31 See endnote 30

32 See http://www.malariavaccine.org/from-the-field.php

33 See video at http://www.gatesfoundation.org/press-releases/Pages/decade-of-vaccines-wec-announcement-100129.aspx

34 See http://www.gavialliance.org/resources/GAVI_Alliance_Evidence_Base_March_2010.pdf

35 See http://www.theglobalfund.org/en/pressreleases/?pr=pr_100608 (press release); http://www.theglobalfund.org/en/malaria/?lang=en

36 Black, R.E. et al. (2008) Maternal and child undernutrition: global and regional exposures and health consequences. *Lancet* 371: 243-260

37 Investing in the Future: A call to action on vitamin and mineral deficiencies, 2009, accessed from: http://www.unitedcalltoaction.org/documents/Investing_in_the_future.pdf)

38 See http://www.unicef.org/rightsite/sowc/pdfs/statistics/SOWC_Spec_Ed_
CRC_TABLE%202.%20NUTRITION_EN_111309.pdf

39 See endnote 38

40 Guiding Principles for complementary feeding of the breastfed child.
WHO Report, accessed from: http://whqlibdoc.who.int/paho/2003/
a85622.pdf; also see http://www.who.int/mediacentre/factsheets/fs342/
en/index.html

41 See endnote 40

42 See http://www.copenhagenconsensus.com/Research/Best%20Practice%20
Papers.aspx

43 Press Release Copenhagen Consensus, 2008—Results. See
http://www.copenhagenconsensus.com/Files/Filer/CC08/Presse%20%20
result/Copenhagen_Consensus_2008_Results_Press_Release.pdf

44 Millennium Development Goals Report, UN, 2010 "Globally, the total number
of under-five deaths declined from 12.5 million in 1990 to 8.8 million in
2008." See http://www.un.org/millenniumgoals/pdf/MDG%20Report%20
2010%20En%20r15%20-low%20res%2020100615%20-.pdf#page=28)

45 Bryce, J. et al. (2008) Maternal and child undernutrition: effective action at
national level. Lancet 371: 510-526

46 Progress on Child and Maternal Nutrition, UNICEF, 2009. See http://www.
unicef.pt/docs/Progress_on_Child_and_Maternal_Nutrition_EN_110309.pdf

47 Investing in the future: a united call to action on vitamin and mineral
deficiencies. Global Report, 2009. UNICEF, accessed: http://www.unitedcall-
toaction.org/documents/Investing_in_the_future.pdf

48 See http://stream.idrc.ca/communications/lastingimpacts/mi_english/

49 See http://www.micronutrient.org/english/View.asp?x=584

50 See http://www.micronutrient.org/CMFiles/MI-AnnualReport0809-EN-
web.pdf (total expenses: 2008-29,297,864 and 2009-29,411,683)

51 Micronutrient Initiative, 2008-2009 Annual Report, accessed: http://www.
micronutrient.org/CMFiles/MI-AnnualReport0809-EN-web.pdf

52 Zlotkin, S.H. et al. (2005) Micronutrient sprinkles to control childhood anaemia. *PLoS Med* 2(1):e1

53 See endnotes 1, 2 and 3

54 Breastfeeding.com. The Nestle boycott. See http://www.breastfeeding.com/advocacy/advocacy_boycott.html.

55 Baker, J.C. (1985) The international infant formula controversy: a dilemma in corporate social responsibility. *Journal of Business Ethics* 4(3):181-190

56 Newton, L.H. (1999) Truth is the daughter of time: the real story of the Nestle Case. *Business and Society Review* 104(4):367-395

57 Editorial (1976) The Infant-food Industry. *The Lancet* 719(513)

58 *Multinational corporations and the impact of public advocacy on corporate strategy: Nestle and the infant formula controversy.* (1994) Kluwer Academic Pub., Boston

59 The International Code of Marketing of Breast Substitutes, WHO, was adopted in May 21, 1981. For details see: http://www.who.int/nutrition/publications/code_english.pdf

60 Newton, L.H. (1999) Truth is the daughter of time: the real story of the Nestle Case. *Business and Society Review* 104(4):367-395

61 See endnote 2

62 Wise, J. (1997) Baby milk companies accused of breaking marketing code. *BMJ* 31(7083):830-831

63 See http://www.gainhealth.org/press-releases/gain-receives-us-38-million-gates-foundation-grant-save-children's-lives-healthy-food

64 See endnote 63

65 Singh et al. (2010) Shared principles of ethics for infant and young child nutrition in the developing world. *BMC Public Health* 10:321

66 See endnote 65

67 The 2008 Summit theme, "The Global Nutrition Challenge," ushered in the Access to Nutrition Initiative, a collaborative project led by the Bill & Melinda Gates Foundation and the Wellcome Trust. This will be the first ranking index of its kind to map the food and beverage industry

worldwide." See http://www.pacifichealthsummit.org/downloads/2009%20 Summit/2009%20Summit%20Report.pdf

68 See http://www.independent.co.uk/news/world/americas/nike-admits-to-mistakes-over-child-labour-631975.html

69 The State of Food Insecurity in the World (2010). Food and Agriculture Organisation of the United Nations. See http://www.fao.org/docrep/013/ i1683e/i1683e.pdf

70 Paarlberg, R. and Borlaug, N. (2008) *Starved for Science*. Boston, MA, Harvard University Press: pg: 163

71 Finegold, D.L. et al. (2005) *Bioindustry Ethics: Monsanto Company Bio-Agricultural Pioneer*. Burlington, MA. Elsevier Academic Press: pg. 278

72 See http://www.imdb.com/title/tt0088247/taglines

73 Finegold, D.L. et al. (2005) *Bioindustry Ethics: Monsanto Company Bio-Agricultural Pioneer*. Burlington, MA. Elsevier Academic Press: pg. 283

74 Paarlberg, R. and Borlaug, N. (2008) *Starved for Science*. Boston, MA, Harvard University Press: pg: 175

75 See endnote 74

76 Mabaye, J. et al. (2010) The Role of Trust Building in the Development of Biosafety Regulations in Kenya', 6/2 *Law, Environment and Development Journal*, available at http://www.lead-journal.org/content/10216.pdf

77 Mabaye, J. et al. (2010) The Role of Trust Building in the Development of Biosafety Regulations in Kenya', 6/2 *Law, Environment and Development Journal*, available at http://www.lead-journal.org/content/10216.pdf: pg. 216

78 See endnote 76

79 Finegold, D.L. et al. (2005) *Bioindustry Ethics: Monsanto Company Bio-Agricultural Pioneer*. Burlington, MA. Elsevier Academic Press

80 See http://monsanto.mediaroom.com/index.php?s=43&item=607

81 See http://www.monsanto.ca/about/pledge/default.asp

82 See http://www.monsanto.com/ourcommitments/Pages/water-efficient-maize-for-africa.aspx

Chapter Six

1 Chakma et al. (2010) Case Study: India's billion dollar biotech. *Nature Biotechnology* 28:783

2 See endnote 1

3 Frew et al. (2009) A business plan to help the 'Global South' in its fight against neglected diseases. *Health Affairs* 28(6):1760-73

4 Hason, A.K. and Shimotake, J.E. (2006) Recent development in patent rights for pharmaceuticals in China and India. *Pace International Law Review* 18(1):303-16

5 Rosenburg, T. (2001) How to Solve the World's AIDS Crisis. *The New York Times Magazine*, Jan 28: 26

6 Frew et al. (2007) India's health biotech sector at a crossroads. *Nature Biotechnology* 25(4): 403-417; Jadhav, S. et al. (2010) Influenza vaccine production capacity building in developing countries: example of the Serum Institute of India. *Procedia in Vaccinology* 2(2):166-171

7 Zakaria, F. (2008) *The post-American world*. New York, N.Y., W.W. Norton & Company: pg. 1

8 See endnote 7, pg. 2

9 See endnote 1

10 See endnote 1

11 Blumberg, B.S. (1997) Hepatitis B virus, the vaccine, and the control of primary cancer of the liver. *PNAS* 94:7121-7125

12 See http://nobelprize.org/nobel_prizes/medicine/laureates/1976/blumberg.html

13 See http://www.hepb.org/professionals/hepatitis_b_vaccine.htm

14 See endnote 1

15 See http://www.who.int/immunization/topics/hib/en/index.html

16 Health Body Cancels Shan5 pre-qualification. See http://www.fiercepharma.com/story/health-body-cancels-shan5-pre-qualification/2010-08-03

17 Frew et al. (2007) India's health biotech sector at a crossroads. *Nature Biotechnology* 25(4): 403-417

18 See endnote 17

19 Forbes dubbed her "Biotech Queen." See http://www.biocon.com/ biocon_press_news_details.asp?subLink=news&Fileid=190

20 Palnitkar, U. (2005) Growth of Indian biotech companies, in the context of the international biotechnology industry. *Jour. Of Comm. Biotech.* 11(2):146-154; See http://in.reuters.com/article/idINIndia-51756420100927

21 Biocon buys Nobex for $5 million, accessed: http://ia.rediff.com/ money/2006/mar/29biocon.htm?q=bp&file=.htm

22 Yang, W.Z. et al. (2010) Safety Evaluation of Allogeneic Umbilical Cord Blood Mononuclear Cell Therapy for Degenerative Conditions. *Journal of Translational Medicine* 8:75

23 Frew et al. (2008) The Indian and Chinese health biotechnology industries: potential champions of global health? *Health Affairs* 27(4):1029-1041

24 See http://www.fda.gov/AboutFDA/CentersOffices/OC/Officeof InternationalPrograms/ucm115256.htm

25 Guyatt, G.H. et al. (1990) The n-of-1 randomized controlled trial: clinical usefulness. Our three-year experience. *Ann Intern Med.* 112(4):293-9.

26 Cyranoski, D. (2009) Stem-cell therapy faces more scrutiny in China. *Nature* 459:146–147.

27 See endnote 26

28 See endnote 23

29 Hason, A.K. and Shimotake, J.E. Recent developments in patent rights for pharmaceuticals in China and India. *Pace Int'l L. Rev.* 18:303-316

30 See endnote 23

31 See endnote 23

32 Pan, J. et al. (2009) Effect of recombinant adenovirus-p53 combined with radiotherapy on long term prognosis of advanced nasopharyngeal carcinoma. *Journal of Clinical Oncology* 27(5):799-804

33 See endnote 32

34 See http://www.pharmaasia.com/article-6144-bendamajorityownerofshenz-
    hensibionogenetech-asia.html and http://www.fiercebiotech.com/node/6381

35 Frew et al. (2008) Chinese health biotech and the billion patient market.
    Health Affairs 26(1): 37-53

36 See http://www.who.int/mediacentre/factsheets/fs328/en/index.html

37 See endnote 35

38 See endnote 3

39 See endnote 3

40 See endnote 3

41 See endnote 3

42 See http://www.who.int/topics/leishmaniasis/en/

43 See endnote 3

44 See http://www.timesonline.co.uk/tol/news/science/medicine/article-
    6994620.ece

45 See http://www.temasekholdings.com.sg/news_room/press_speeches/
    BioVeda%20China%20Press%20Release%20-final-2Jun05.pdf;
    http://www.bioveda.com.cn/en/

Chapter Seven

1   See http://www.ippmedia.com/frontend/index.php?l=13978

2   See http://www.afdb.org/fileadmin/uploads/afdb/Documents/
    Publications/27842402-EN-HLP-REPORT-INVESTING-IN-AFRICAS-
    FUTURE.PDF

3   See http://www.ifc.org/ifcext/media.nsf/content/SelectedPressRelease?
    OpenDocument&UNID=E9EA2F46CC97CA84852573B5005DA423

4   See http://www.kigalimemorialcentre.org/old/index.html

5   See http://www.usp.ac.fj/worldbank2009/frame/Documents/Publications_
    regional/BuildingSTICapacityinRwanda.pdf

Chapter Eight

1   World Health Organization 10 Facts on Malaria, March, 2009, accessed: http://www.who.int/features/factfiles/malaria/en/index.html

2   See endnote 1

3   Alonso, P.L. (2011) A research agenda to underpin malaria eradication. 8(1):e1000406; World Malaria Report (2010), WHO Global Malaria Programme, accessed http://www.who.int/malaria/world_malaria_report_2010/en/index.html

4   World Malaria Report (2010), WHO Global Malaria Programme, accessed http://www.who.int/malaria/world_malaria_report_2010/en/index.html; Flaxman, A.D. (2010) Rapid Scaling Up of Insecticide-Treated Bed Net Coverage in Africa and Its Relationship with Development Assistance for Health: A Systematic Synthesis of Supply, Distribution, and Household Survey 7(8):e100328, accessed: http://www.plosmedicine.org/article/info:doi/10.1371/journal.pmed.1000328

5   Roll Back Malaria Progress & Impact Series—Focus on Senegal, Number 4-November 2010, World Health Organization, accessed: http://www.rollbackmalaria.org/ProgressImpactSeries/docs/report4-en.pdf

6   Mutabingwa, T.K. Artemisinin-based combination therapies (ACTs): best hope for malaria treatment but inaccessible to the needy! *Acta Trop.* 2005 Sep;95(3):305-15.

7   Ro, D.K. et al. (2006) Production of the antimalarial drug precursor artemisinic acid in engineered yeast. *Nature* 440:940-943.; Ro, D.K. et al. (2008) Induction of multiple pleiotropic drug resistance genes in yeast engineered to produce an increased level of anti-malarial drug precursor, artemisinic acid. BMC *Biotechnol.* 8:83.; Hommel, M. The future of artemisinins: natural, synthetic or recombinant? J *Biol.* :7(10):38.

8   Fact Sheet: RTS,S Malaria Vaccine Candidate, GlaxoSmithKline Biologicals and Malaria Vaccine Initiative, accessed: http://www.worldmalariaday.org/download/partners/Updated_RTSS_FactSheet_21_April_2010.pdf

9 Initiative for Vaccine Research. Most advanced malaria vaccine candidate and timing for policy recommendations. World Health Organization, accessed: http://www.who.int/vaccine_research/diseases/malaria/vaccine_candidate_policy/en/index.html; Olotu, A. et al. (2011) Efficacy of RTS,S/AS01E malaria vaccine and exploratory analysis on anti-circumsporozoite antibody titres and protection in children aged 5–17 months in Kenya and Tanzania: a randomised controlled trial. *The Lancet Infectious Diseases* 11(2):102-109)

10 Hoffman, S.L. et al. (2010) Development of a metabolically active, non-replicating sporozoite vaccine to prevent Plasmodium falciparum malaria. *Hum Vaccin.* 6(1):97-106.

11 Rerks-Ngarm, S. et al. (2009) Vaccination with ALVAC and AIDSVAX to prevent HIV-1 infection in Thailand. *New England Journal of Medicine* 361: 2209-2220

12 RV144 in Detail, accessed: http://www.iavireport.org/publicationsand-graphics/Documents/RV144_InDetail.pdf

13 NIH-Led scientists find antibodies that prevent most HIV strains from infecting human cells. *NIH News*, July 8, 2010, accessed: http://www.nih.gov/news/health/jul2010/niaid-08.htm; Wu, X. et al. (2010) Rational design of envelope surface identifies broadly neutralizing human mono-clonal antibodies to HIV-1. *Science* 329(5993):856-861; Zhou, T. et al. (2010) Structural basis for broad and potent neutralization of HIV-1 by antibody VRC01. *Science* 329(5993):811-817

14 HIV antibody finding may be a new piece to AIDS puzzle, *American Scientist*, accessed: http://www.americanscientist.org/science/science.aspx?id=10122&content=true

15 Auyert, B. et al. (2005) Randomized, controlled intervention trial of male circumcision for reduction of HIV infection risk: the ANRS 1265 Trial. *PLoS Med.* 2(11):e298.; Bailey, R.C. et al. (2007) Male circumcision for HIV prevention in young men in Kisumu, Kenya: a randomised controlled trial. *Lancet* 369(9562), 643-656; Gray, R.H. et al. (2007) Male circumcision for HIV prevention in men in Rakai, Uganda: a randomised trial. *Lancet* 369(9562): 657-666

16 Africa: tracking the male circumcision rollout. IRIN, Humanitarian News and Analysis, UN Office, March 2, 2010, accessed: http://www.irinnews. org/report.aspx?ReportID=88286; Clearinghouse on male Circumcision for HIV Prevention. Tools and Guidelines, accessed: http://www.male-circumcision.org/programs/tools_guidelines.html

17 Weiss, H.A. (2008) Male circumcision for HIV prevention from evidence to action? AIDS 22(5):567-74

18 For more information on the PrEP Trials visit the following site: http://www.avac.org/ht/d/sp/i/3507/pid/3507

19 Grant, R.M. (2010) Preexposure Chemoprophylaxis for HIV prevention in men who have sex with men. NEJM 363:2587-2599

20 2010/2011 Tuberculosis Global Facts, WHO, accessed: http://www.who.int/ tb/publications/2010/factsheet_tb_2010_rev21feb11.pdf

21 Towards universal access to diagnosis and treatment of multidrug-resistant and extensively drug-resistant tuberculosis by 2015, WHO, accessed: http://www.who.int/tb/features_archive/world_tb_day_mdr_ report_2011/en/index.html

22 Ma, Z. (2010) Global tuberculosis drug development pipeline: the need and the reality. Lancet 375(9731):2100-2109

23 See endnote 22

24 To learn more, see: http://www.tballiance.org/cptr/

25 FDA awards nearly $3 million for TB research, TB Alliance News Brief, 2010, accessed: http://www.tballiance.org/newscenter/view-brief.php?id=939

26 TB Alliance launches first clinical trial of a novel TB drug regimen, TB Alliance Press Release, 2010, accessed: http://www.tballiance.org/ newscenter/view-brief.php?id=942

27 Aagaard, C. et al. (2011) A multistage tuberculosis vaccine that confers efficient protection before and after exposure. Nature Medicine 17: 189-194

28 Boehme, C.C. et al. (2010) Rapid molecular detection of tuberculosis and

rifampin resistance 363(11):1005-1015; see also Cepheid website: http://www.cepheid.com/company/news-events/in-the-news/

29 This price was negotiated by FIND and represents a significant reduction when compared to the manufacturer's list prices in Europe and the USA." Source: http://www.finddiagnostics.org/media/news/110221.html. For a full breakdown of pricing model, see: http://www.finddiagnostics.org/programs/tb/find-negotiated-prices/xpert_mtb_rif.html and http://www.who.int/tb/laboratory/roadmap_xpert_mtb-rif.pdf

30 Bill and Melinda Gates pledge $10 billion in call for decade of vaccines, January 29, 2010, accessed: http://www.gatesfoundation.org/press-releases/Pages/decade-of-vaccines-wec-announcement-100129.aspx

31 2011 Annual letter from Bill Gates, accessed: http://www.gatesfoundation.org/annual-letter/2011/Documents/2011-annual-letter.pdf

32 Dugger, C. New meningitis vaccine brings hope of taming a ravaging illness in Africa. New York Times, December 04, 2010, accessed, http://www.nytimes.com/2010/12/05/world/africa/05meningitis.html?_r=1

33 See endnote 32

34 Oshinsky, D.M. (2005) Polio: an American story. Oxford University Press: pg. 161

35 Forty-first World Health Assembly, Geneva, 2-13 May, 1988: WHA41.28 Global eradication of poliomyelitis by the year 2000, accessed: http://www.who.int/csr/ihr/polioresolution4128en.pdf; Polio Global Eradication Initiative: Progress towards polio eradication, accessed: http://www.polioeradication.org/Aboutus/Progress/Progresstowardspolioeradication.aspx

36 World Health Organization, Poliomyelitis, Fact sheet No. 114, November 2010, accessed: http://www.who.int/mediacentre/factsheets/fs114/en/index.html

37 See endnote 36

38 Bill's Annual Letter: End Polio Now, The Gates Notes, February 15, 2011, accessed: http://www.thegatesnotes.com/Thinking/Bills-Annual-Letter-End-Polio-Now

39 See endnote 38

40 Collins, F.S. (2010) *The language of life: DNA and the revolution in personalized medicine*. New York, N.Y., Harper Collins Publishers: pg. 85

41 Daar, A.S. Vision of a personal genomics future. *Nature* 463:298-299

42 Pre-announcement of a request for applications on implementation research on hypertension in low and middle income countries, Global Alliance for Chronic Diseases, December 8, 2010, accessed: http://www.ga-cd.org/pdf/GACD%20Pre-announcement%20Hypertension_final.pdf

43 To learn more, go to: The Global Task Force on Expanded Access to Cancer Care and Control in Developing Countries, at: http://gtfccc.harvard.edu/icb/icb.do?keyword=k69586&pageid=icb.page384726

44 Juma, C. (2010) *The new harvest: agricultural innovation in Africa*. Oxford University Press

45 Immelt, J. et al. (2009) How GE is disrupting itself. *Harvard Business Review*, accessed http://hbr.org/2009/10/how-ge-is-disrupting-itself/ar/1

46 TED Talks, TEDIndia, 2009: Filmed Nov, 2009; Posted December 2009, accessed: http://www.ted.com/talks/thulasiraj_ravilla_how_low_cost_eye_care_can_be_world_class.html

47 Canada, embrace this Grand Challenge. Editorial, *Globe and Mail*, Wednesday, March 09, 2011, accessed http://www.theglobeandmail.com/news/opinions/editorials/canada-embrace-this-grand-challenge/article1934332/

48 Masum, H. et al. (2010) Africa's largest long-lasting insecticide-treated net producer: lessons from A to Z Textiles. *BMC International Health and Human Rights*

49 McNeil Jr., D.G. Five years in, gauging impact of Gates grants. *The New York Times*, December 20, 2010, accessed: http://www.nytimes.com/2010/12/21/health/21gates.html

50 Rafael, M.E. et al. (2006) Reducing the burden of childhood malaria in Africa: the role of improved diagnostics *Nature* 444(Suppl 1):39-48

51 Saving Lives at Birth Press Release, Grand Challenge for Development, March 9, 2011, accessed: http://savinglivesatbirth.net/news/11/03/09/saving-lives-birth-press-release

52 Saving Lives at Birth Presentation, Grand challenge for development, accessed http://savinglivesatbirth.net/news

53 See endnote 52

54 Grantham-McGregor, S. et al. (2007) Development potential in the first 5 years for children in developing countries. *Lancet* 369:60-70

55 To learn more, read: Singer, P., Daar, A., Dowdeswell, E. (2003) Bridging the Genomics Divide. *Global Governance* 9(1):1-6 ; Acharya, T., Daar, A., Thorsteinsdottir, H., Dowdeswell, E., Singer, P. (2004) Strengthening the Role of Genomics in Global Health. *PLoS Med*, 1(3):e40; Dowdeswell, E., Daar, A., Singer, P. (2005) Getting governance into genomics. *Science and Public Policy* 23(6):497

56 Rischard, J.F. (2001) High Noon: We need new approaches to global problem-solving, fast. *Journal of International Economic Law* 4(3):507-525

57 See the discussion at Carnegie Council, A New World Order, Public Affairs Program, Slaughter, A.M. and Myers, J.J. April 15, 2004, accessed http://www.cceia.org/resources/transcripts/4467.html

58 Calestous Juma testifies before congress on US/African Relations. Testimony before the House Committee on Foreign Affairs. Subcommittee on Africa and Glboal health. United States House of Representatives. July 18, 2007, accessed http://www.hks.harvard.edu/news-events/news/testimonies/calestous-juma-testifies-before-congress-on-us-african-relations

59 Sachs, J. (2005) *The end of poverty.* Penguin Group USA Inc. New York, New York: pgs 1-3

60 See endnote 59

61 See endnote 59

62 See endnote 59

63 Moyo, D. (2010) *Dead aid.* New York, N.Y. Farrar, Straus and Giroux

64 See endnote 31

65 Zakaria, F. (2008) *The post American world.* New York, N.Y., WW Norton & Co.

ACKNOWLEDGEMENTS

This book describes our work over the past decade or so. We worked with a large number of colleagues, research fellows, assistants, associates, graduate and undergraduate students, and many, many supporters.

We were lucky to have the support and encouragement of Sandra and Joseph Rotman, friends and benefactors, for over a decade. Joseph L. Rotman has in many ways been the third member of our duo, and he has significantly inspired, supported, shaped, and mentored our journey.

Robert Bell, CEO of University Health Network, and Catharine Whiteside, Dean of Medicine at University of Toronto, have been steadfast supporters. David Naylor, President of the University of Toronto, and Ilse Treurnicht, CEO of MaRS, have been helpful and supportive in many ways for over a decade. Some great Canadians, including John Evans, Cal Stiller, and Henry Friesen have been encouraging and supportive from the beginning and at many points along the way.

Many of our distinguished colleagues are the people readers meet in this book. We will not repeat all their names here. We would be remiss, however, not to single out Harold Varmus, who led the Gates

Foundation Grand Challenges Scientific Advisory Board at its inception; John Bell, who chaired the Oxford Health Alliance and continues to be a great supporter of the Global Alliance for Chronic Diseases; and Calestous Juma, of Harvard University, who has been a great muse for us. All the members of the scientific and governing boards of the Grand Challenges in Global Health, the Global Alliance for Chronic Disease, Grand Challenges Canada, and the McLaughlin-Rotman Centre for Global Health have been incredible advisors and mentors.

We have benefited a great deal from more than a decade's association with Elizabeth Dowdeswell, Jeff Sturchio, Dan Carucci, Alan Bernstein, and Derek Yach. In our work on African innovation, we have worked closely with Minister Peter Msolla, Professor Evelyne Mbede, Dr. Mwele Malecela, Professor Wen Kilama, Dr. Hassan Mshinda, Professor Burton Mwamila in Tanzania; Professor Nelson Sewankambo in Uganda; Minister Romain Murenzi (who has recently taken over from yet another of our wise colleagues and advisors, Dr. Mohammed Hassan, as CEO of TWAS), Mike Hughes, and Christine Gasingirwa in Rwanda; John McDermott in Kenya; and the late Minister Major Courage Quashiga, and Francis Nkrumah, in Ghana.

A huge source of inspiration and help came from our remarkable students, assistants, research assistants and associates, consultants, and colleagues at the McLaughlin-Rotman Centre for Global Health (and earlier at the U of T Joint Centre for Bioethics), many of whom the reader also meets in the book. They include: Halla Thorsteinsdottir, Tara Acharya, Sarah Frew, Ross Upshur, Jim Lavery, Jerome Singh, Anant Bhan, Paulina Tindana, Sunita Bandewar, Claudia Emerson, Shane Green, Lauren Leahy, Lara El-Zahabi, Arisa Goldstone, Obidimma Ezekia, Jennifer Deadman, Justin Mabeya, Rahim Rezaie, Beatrice Seguin, Sarah Ali-Khan, Hassan Masum, Jennifer Heys, Justin Chakma, Dominique McMahon, Ken Simiyu, Sheila Kamunyori, Ronak Shah,

Monali Ray, Deepa Persad, Fabio Salamanca-Buentello, Katie Berndtson, Fiona Thomas, Jee Yon Kim, Michael Keating, Sina Zere, Mehrdad Hariri, Munira Tayabali, Tijana Markovich, Leticia Law, Heather Navarra, David Brook, and one who has tragically passed away, the incomparable and universally loved Sara Al Bader. (There are others beyond this list whom we have not specifically named but do want to thank.) We also want to acknowledge our wonderful new colleagues at Grand Challenges Canada, and the incredible Andrew Taylor who has been working with us for this whole decade; without him this work would not have been possible. Billie-Jo Hardy deserves special mention: Thank you Billie for your incredible and diligent fact-checking of the manuscript for this book. In many parts of our journey, we are drawing upon our collective work with these amazing people, and this book is a tribute to their work.

We are grateful to our terrific literary agent, Bruce Westwood, and his son Ashton Westwood, from Westwood Agency, and the remarkable John Fraser for introducing us. Michael Keating helped to edit the early manuscript and later Sarah Scott helped us through a major re-edit and re-writing. At Doubleday/Random House, we have encountered an incredible amount of support, encouragement, and practical help from Kristin Cochrane, Maya Mavjee, Martha Kanya-Forstner, Nita Pronovost and, above all, our amazing, warm, and wonderful final editor, Lynn Henry.

We thank the Canadian Institutes of Health Research, Genome Canada, Ontario Genomics Institute, Ontario Research Fund, Rockefeller Foundation, and the Bill & Melinda Gates Foundation for supporting our work through research grants.

Bill and Melinda Gates are a huge inspiration to everyone interested in global health. At the Bill & Melinda Gates Foundation we want to specifically thank those with whom we have worked closely for almost the whole decade: Tachi Yamada, Carol Dahl, Fil Randazzo, and Steve Buchsbaum, along with Josie Ekberg and Kristi Anthony. We have learnt a huge amount from them and their amazing colleagues—too numerous to list here—at the foundation. In Canada, David Malone, Rohinton Medhora, and Michael Clarke at IDRC, and Alain Beaudet, Pierre Chartrand, and Christine Fitzgerald at CIHR have equally been great supporters and wise guides.

This book is based on our documented joint research, which we have done collectively with many students, assistants, research assistants and associates, consultants, and colleagues as highlighted in these pages and above. We are deeply grateful to them. When we say "we" in this book, we usually mean the duo of Peter and Abdallah, although sometimes—and we have tried to so indicate—we mean one of us alone. Rarely, the "we" refers to our research team where neither of us was present but the encounter involved research we supervised. We have used not only our research but also interviews conducted specifically for this book to re-establish particulars of events and experiences of others. We have included many personal recollections based on our experiences over the past dozen years; these are necessarily subject to our memories but they add an important first-person flavour to the book. The facts herein have been carefully checked, but we apologise in advance for any errors and omissions.

We also want to acknowledge each other. An academic relationship like ours, lasting over a dozen years, is highly unusual and a testament to our friendship and commitment to a common cause. We want to acknowledge all those who work for that cause—all those committed to improving the lives of billions of people through global health innovation.

And, of course, we want to acknowledge our families, to whom this book is dedicated.

INDEX